Geoheritage, Geoparks and Geotourism

Conservation and Management Series

Spectacular geo-morphological landscapes and regions with special geological features or mining sites, are becoming increasingly recognized as critical areas to protect and conserve for the unique geoscientific aspects they represent and as places to enjoy and learn about the science and history of our planet. More and more national and international stakeholders are engaged in projects related to “Geoheritage”, “Geo-conservation”, “Geoparks” and “Geo-tourism” and are positively influencing the general perception of modern Earth sciences. Most notably, “Geoparks”, have proven to be excellent tools to educate the public about “Earth Sciences”. And shown to be areas for recreation and significant sustainable economic development through geotourism. In order to develop further the understanding of earth sciences in general and to elucidate the importance of earth sciences for Society the Geoheritage, Geoparks and Geotourism Conservation and Management Series has been launched together with its sister GeoGuides series. “Projects” developed in partnership with UNESCO, World Heritage and Global Geoparks Networks, IUGS and IGU, as well as with the “Earth Science Matters” Foundation, are welcome. The series aims to provide a place for in-depth presentations of developmental and management issues related to Geoheritage and Geotourism as well existing and potential Geoparks. Individually authored monographs as well as edited volumes and conference proceedings are welcome in this series. This book series is considered to be complementary to the Springer-Journal “Geoheritage”.

More information about this series at http://www.springer.com/series/11639

Jiun-Chuan Lin • Shew-Jiuan Su

Geoparks of Taiwan

Their Development and Prospects for a Sustainable Future

Springer

Jiun-Chuan Lin
Department of Geography
National Taiwan University
Taipei, Taiwan

Shew-Jiuan Su
Department of Geography
National Taiwan Normal University
Taipei, Taiwan

ISSN 2363-765X ISSN 2363-7668 (electronic)
Geoheritage, Geoparks and Geotourism
ISBN 978-3-030-04893-8 ISBN 978-3-030-04895-2 (eBook)
https://doi.org/10.1007/978-3-030-04895-2

Library of Congress Control Number: 2018963996

This Springer imprint is published by the registered company Springer Nature Switzerland AG
The registered company address is: Gewerbestrasse 11, 6330 Cham, Switzerland

Foreword I

Earth science communication can be difficult, with explanations of complex geodynamic underground or surface processes and the use of scientific terminology often confusing those we want to take with us on a journey to understand and respect our restless planet. While science communication in general can be demanding, it is a prerequisite to convince people, decision-makers, the public at large, by providing up-to-date scientific knowledge on how to maintain and develop our environment in a sustainable way and sometimes to reduce mistrust towards science experts, as well as partly a bias towards industry and tourism.

To cut a long story short: A wonderful tool to approach these challenging goals with respect to Earth sciences is a 'Geopark'.

Nearly 30 years ago, a few geologists, palaeontologists, geographers and environmental planners, sometimes recognised as 'geo-enthusiasts', recognised that conservation and development of significant geological and geomorphic sites and landscapes could provide a fundamental educational tool in our quest for understanding Earth, planets and life, as well as providing a promotional tool for sustainable regional development through 'geotourism'.

The 1984 inauguration of the '*Parc Géologique de Haute Provence*' in France, initiated by Guy Martini, set the stage for many followers. Establishing geo-related inventories and assessing the geoscientific value of sites started in the late 1980s and under the leadership of John Cowie (UK), geoscientists of the UNESCO-IUGS-IGCP community began in 1989 to compile and identify, country by country, the most important sites, in order to establish a '*Global Indicative List of Geological Sites*' (GILGES).

The earliest use of the term 'Geopark' dates also back to 1989 when the terms 'Geo-Pfad' (Geo-Trail) and 'Geo-Park' were used in Germany as synonyms by geologists around Marie-Luise Frey and planners from Gerolstein/Eifel. 1993 the Austrian geologists Lutz H. Kreutzer and Hans-Peter Schönlaub proposed the mountain 'Wendelstein' in the Bavarian Alps as a 'Geo-Park', in same year the first 'real' Geopark in Germany was established in the Eifel region and 1994 followed the Japanese 'Itoigawa Geopark'.

The highly ambitious and commendable challenge of Cowie to compile a word-wide list of the most important geosites remained incomplete; it was later taken up under the guidance of Bill Wimbledon by the then active IUGS-Working Group '*Geosites*', aiming mainly at conservation, and a successor project of the '*European Association for the Conservation of Geological Heritage*' (ProGEO).

Apart from these inventory works, a significant global societal breakthrough in Geoheritage was reached through the 1991 '*Digne Declaration on the Rights of the Memory of the Earth*'. ProGEO's Working Group, under the leadership of Guy Martini, highlighted the global value of geology, geodiversity and geoconservation for the society.

Taking all these activities and undertakings into account, UNESCO's former Division of Earth Sciences, under my directorship, explored since 1994 the possibilities to create a global network of selected, geoheritage-related territories by offering its international umbrella for the numerous existing, but diverse national efforts in promoting the knowledge of the Earth. At the 30th International Geological Congress (Beijing 1996), a UNESCO Workshop on 'Geological Heritage and Geosites/Geoparks' was organized by me and Mechtild Rössler (UNESCO,

World Heritage Centre), Paul Dingwall (New Zealand) and Zhao Xun (China, Secretary General of the 30th IGC): An innovative—but sometimes perceived as too popular and 'unscientific'—initiative was therefore on its way.

Today, *geoparks*, as a new international 'brand', are best defined through the criteria and guidelines of the *UNESCO Global Geoparks* designation that forms since November 2015 one part of UNESCO's *International Geoscience and Geoparks Programme (IGGP).* One has to mention that in contrast to UNESCO's World Heritage and MAB Programmes, or other agreements like the UN-Convention on Biological Diversity or the Ramsar Convention, UNESCO Global Geoparks are not subject to an international convention, they are policy based under the General Conference of UNESCO.

One of the general principles of the Geoparks concept is, apart from education and conservation, the focus on sustainable development, including sites that represent landscape elements rather than small geological outcrops. This landscape approach is integrating biotic and abiotic Earth heritage conservation and underpins that geoconservation and geological heritage implicitly express the importance of earth history to our cultural heritage.

Managers and earth scientists engaged in geoparks are strengthening dialogue with planners, economists, and many others, in seeking to sustain and develop the world's life-support systems for the benefit of present and future generations. Agriculture and forestry depend on soil conditions, drinking water from the subsurface is our most vital commodity, the industry relies on energy and mineral resources, and last but not least, people seek recreation in the great outdoors.

National and (internationally recognised) global geoparks provide important lessons by touring to high rising cliffs and deep caves and the treasure troves of the earth. Each of the geoparks is different, like the landscape they represent, each opens a window to new exciting experiences in the world around us or below our feet. Thus, the geoparks are contributing to today's huge challenges 'Climate Change', 'Disaster Risk Reduction', 'Clean Water', 'Sustainable Energy' and 'Health'.

Also in Taiwan spectacular geomorphologic landscapes, partly representing catastrophic events (volcanic eruptions, remnants of earthquakes or rock falls), and regions with special erosional and sedimentological features, quarries or engineering and mining sites are becoming increasingly recognised as critical areas to conserve the unique geoscientific aspects they represent as well as places to enjoy and learn about the nature, science and history of our planet and humankind.

Treasures of Taiwanese nature are presented by the Penghu Marine Geopark (with its basaltic floods), the Northern Coast Yehliu Geopark (with spectacular erosional features), the Yenchao Mud Volcano Geopark (with its mud volcanoes and badlands), the Lichi Badland Geopark (with its 'two mélange' landscape), the Eastern Coast Geopark (with its corals and coastal rock features), the Tsaoling Geopark—Caoling Geopark (with its landslides), the Matsu Geopark (with its granite and 'cold war heritage' hidden tunnels), the Southwest Coastal Geopark (with its sand dunes and wetlands), or the Northeast Coastal Longdong and Bitou Cape Geopark (with its marine hot springs).

I was privileged to visit most of these marvellous places and am still overwhelmed by the way in which the group around Jiun-Chuan Lin, Wen-Cheng Wang and Shew-Jiuan Su (all from Taipei) has so impressively adopted and implemented the geopark initiative and philosophy to their country. Since 1999 Jiun-Chuan Lin and his colleagues contacted UNESCO and its former Division of Earth Sciences, they contributed to numerous conferences of the European and Asian Pacific Geopark Networks and have built up fruitful partnerships, inter alia with Japan.

The first idea of compiling a book on the Taiwanese Geoparks was born, as far as I remember, 2011 at an EGN conference organized by the Geopark 'GeoNorvegica' in Norway. In the following years, a network of exciting well-structured geoparks with breathtaking sceneries was build up and since 2017 a national Geoparks Association supports an intense communication nationally, but with international partners, too.

Nine National Geoparks exist now all over Taiwan, with none of them having currently a chance to be certified as UNESCO Global Geopark—despite the well-known fact that geology does not know boundaries.

I certainly hope that this book will have a double effect: firstly, increase the public motivation within Taiwan to explore the intimate integration of biotic and abiotic features with the country's history and culture and, secondly, make the international community curious to gain more and direct information on the natural treasures of this island(s). Last but not least, I optimistically wish that—based on cooperation and communication with the international Science community—the dilemma of the Taiwanese Geoparks being caught between a great mission and rigid UN rules might be overcome in the very near future.

Göttingen, Germany
2018

Wolfgang Eder
Geoscience Centre, University of Göttingen

Foreword II

A *geopark* is an area characterised by a geological framework that provides a significant natural heritage of national and even global significance. Geopark administrations seek to look after and use their heritage asset to promote awareness of key societal issues including conservation problems, environmental protection, education, social well-being and health, geological *hazards* and the development of *disaster* mitigation strategies. Local communities are encouraged to use geopark assets to demonstrate the use of renewable energy and the value of '*green tourism*'. The industry is assisted to use sustainable tourism models which also enhance the geographical character of a place. In many cases, the areas chosen are economically difficult due to terrain difficulties, adverse soil and agricultural conditions or historical industrial legacies. The geopark concept is then used to re-invigorate the economy of the area. In achieving this considerable tourist, education and health benefits are brought to the wider community who become aware of important local products, practices and traditions. This promotes respect for the people, environment and the integrity of the *landscape*. For these reasons, geoparks are often protected under local, regional or national legislation.

The Geopark concept is part of the UNESCO World Heritage programme which has established an important regional and international network. It is not possible for Taiwan to be part of this system. Despite this, Taiwanese people have followed international developments with great interest and have adopted a laudable approach to conservation practice. Their country is very beautiful (Formosa—the Beautiful Island). It is tectonically active, has important evidence of the geological evolution of the subduction margins between the Philippine plate and the Eurasian plate, extraordinary sedimentary sequences and spectacular coastal, raised beach and erosional processes. The history of land use is a fascinating story of the association between traditional practice and difficult soils.

Above all the people of Taiwan have a deep respect for the environment, and it is written into the constitution that the countryside should be respected and conserved. Environmental education is of the highest priority. In response to the international situation, Taiwan has recognised that it has a responsibility to the global community to follow UNESCO guidelines and develop National Parks and Geoparks and look after their land to the highest World Heritage standards. They regard this responsibility as a 'gift' to all people on the planet much in the same way as the Maori people gifted their sacred Ngauruhoe National Park to the people of New Zealand. This is a statement that we should all applaud for it demonstrates how advanced, gifted and responsible the people of Taiwan are. I therefore recommend this book and congratulate the author. It is a very clear and straightforward account of a remarkable story.

London, UK

Prof. Denys Brunsden OBE, DSc, FKC
Emeritus Professor, King's College London

Preface

It has been more than six years since the inception of the idea of writing a book on the geoparks of Taiwan. Throughout the past six years, the geoparks of Taiwan have evolved rapidly and some even dramatically for the better. As a non-UN member state and an island state located on the ring of fire in the Pacific Ocean, Taiwan has a geological foundation that is worth much attention scientifically and educationally. Furthermore, its geological setting, along with the country's subtropical climate and weather, has conditioned the life of the islanders. Vice versa, the islanders, with technological change and policy making, shape the nature of the island state. Such interaction of the human society and the natural setting has become a venue of learning how we could live peacefully and sustainably with the nature. This is exactly what the geoparks have been promoted in Taiwan.

There were changes in the past six years that astonished us very much. For instance, a national legislative amendment of the *Cultural Conservation Act* for geoparks and their development was passed without much political endeavor or maneuver. We take it as a sign of the mature Taiwan civil society for the conservation of its own environment. It also shows Taiwan's social civilization to care about the global initiative for protecting our only planet with geoparks and act for it. Another example of amazing matter is school education through geoparks. Many school kids of various geopark communities have been educated and trained to become interpreters of the sites of their own geopark. This is an embedded act that we could do for the best interest of the future generations.

While education has been one shining aspect of Taiwan's geoparks, there are strong support from community members. The people living around and in the geopark areas become the guarding angels of the environment. With proper understanding of their own environment, the people could live with sustainable ideas and yield better outcomes for either communal purpose or individual benefits. Geoparks in Taiwan are not only about geology, but also about geographies and the people and society in the geographies.

Derived from the UNESCO, Taiwan's geopark promotion has been following the guidelines of the UNESCO global geoparks. With certain adjustment to fit the local situations, Taiwan geoparks are right on the track for a sustainable future. However, geoparks and their development have not been always easy and successful for Taiwan. The NGOs and the scholarly community would have constantly to locate ways for development. The community people of the geoparks would have to adjust throughout. In particular, when new or alternative ideas arise, people need to think and consider if and how adaption or adjustment could be implemented with desirable results. Continued learning and sharing becomes a typical way of life for the geopark community. Taiwan society has been very courageous in the face of adjusting for the better. People would stop, ponder and make decision collectively or individually and act for a better result. For the aging society of Taiwan, geoparks become a way of connecting people to be young and energetic again in front of the nature. This is exactly what a sustainable environment should be and it is a way of contributing to the global society.

Without the support and help of many people and institutes, this volume would not be possible. We would like to firstly express our gratitude to Dr. Wolfgang Eder for his initiative and encouragement for the making of this volume. Dr. Eder's encouragement through his participation of some geopark activities is important to us. We remember the night in the remote geopark of Tsao-ling. A group of community people from the southwest badland area conducted awkward English to greet Dr. Eder by saying "We are good people from the bad land geopark". That was an unforgettable moment. It also revealed how critical of the force of bottom-up mechanism for geoparks.

During the writing of this volume, we received a lot of support and help. The Springer editorial team has been a constant and strong support. We appreciate all and will never thank them enough. The generous support of the Taiwan Forestry Bureau, the Tourism Bureau, and the National Park Service and their excellent and capable civic servants are tremendously important. We would also like to thank friends in the academic circle, in particular Profs. Wen-Cheng Wang of NTNU, Ying-San Liou of NDU, Chien-Tang Lee of NTU and Shyi-Jen Chyi of NKNU. They have been colleagues and comrades in the research for and promoting Taiwan's geoparks for the past decade. In addition, Prof. Denys Brunsden of King's College London has been a great mentor and editorial support. He literally walked through with us for clarifying points and ideas. We appreciate it and owe him greatly. Lastly but not the least are the people of the geopark community. They are the critical pillars for us. To all of them and many other unnamed persons, we owe a debt of gratitude.

Taipei, Taiwan
2018

Jiun-Chuan Lin
Shew-Jiuan Su

Contents

List of Figures

List of Tables

Taiwan's Geoparks

1

1.1 The Development of Taiwan's Geoparks

United Nations Educational, Scientific and Cultural Organization (UNESCO), authorized by the General Assembly of the United Nations, proposed a project "to form a global network to conserve sites of specially or uniquely geological and geographical value" in November, 1999. This project integrates the result of many national or international landscape conservation, such as "Geotopes," "Geosites," and some so-called geological heritages. These sites are representative, unique, irreplaceable, and irreversible in character. They form a baseline for landscape conservation with a particular social value, where community participation and local sustainability are of concern. It is meanwhile framed as geopark in the global community. Experts from different fields around the world hold different understandings and perspectives of the term "Geopark". Some refer to 'geo'- as geology, others as geography and still others pertain it to 'Gaia', the Greek word for Earth (Su and Lin 2014), all of which are included in the four core values of geoparks. The most tangible and within the Geoparks-community adopted definition stems from Chris Woodley Stuart (UK) "Geoparks are not about rocks, they are fundamentally about people".

According to UNESCO, the main purpose of Geoparks is to reach environmental conservation, to enhance regional social economic development, as well as to integrate the natural and social environment in order to attain sustainability. By raising public awareness of the value of the Earth heritage and the knowledge of environmental carrying capacity, we get to make wiser use of natural resources and strike a balance between humankind and the environment especially in the island state of Taiwan.

The landscapes of Taiwan are unique, from the alpine peaks to the coast, from the countryside to the city. The solid foundation for Taiwan's geoparks is the inventory surveyed, recorded and revised constantly by the Forestry Bureau, Council of Agriculture. The current inventory of 341 geo-sites is a result of decades of survey and revision due to the dynamically constant changing landforms of Taiwan (Lin 2011).

Starting in 2011 the Forestry Bureau supported four geopark communities to observe and watch their local environments. The communities then networked to form an alliance to share and learn from each other. The four are: Penghu Marine Geopark, Tsaoling Geopark, Lichi Badland Geopark, and Yenchao Mud Volcano Geopark. Later the Tourism Bureau promoted two other geoparks, Northern Coast Yehliu Geopark and Matsu Geopark, Taiwan had thus a total of six geoparks in 2011.

The geoparks are strategically designated by Taiwan's academic society to probe how communities might be capable of developing geo-tourism, to conduct landscape conservation and to develop sustainable local social economies. By establishing geoparks in Taiwan, we not only hope to conserve the unique and valuable geo-heritage and landscape of Taiwan, but also anticipate that environmental education can be deeply rooted into our national education and life-long learning systems through geopark practice (Lin 2014; Chen et al. 2012 and Chen 2014). All these may pave an avenue to make geo-tourism environmentally responsible, so to live up to the aspiration of eco-tourism. Sustainable geo-tourism with local participation in the conservation of landscape may strengthen or even create a sense of home affiliation for the locals. With the local participation, local pride and local wisdom, a much valued bottom-up mechanism can be established in order to guard the local environment for living quality and to steward valuable geo-heritage sites for showing them to visitors. Thus, local social and economic development can be integrated into an environmental context and consequently create a unique socio-economy. Each selected region can discover and decide about its own significant geological sites that represent local and regional values that enable them to contribute to the National Comprehensive Development Plan and/or the County/City Comprehensive Developmental Plan. Along this line successfull Geoparks could be established that help

J.-C. Lin and S.-J. Su, *Geoparks of Taiwan*, Geoheritage, Geoparks and Geotourism,
https://doi.org/10.1007/978-3-030-04895-2_1

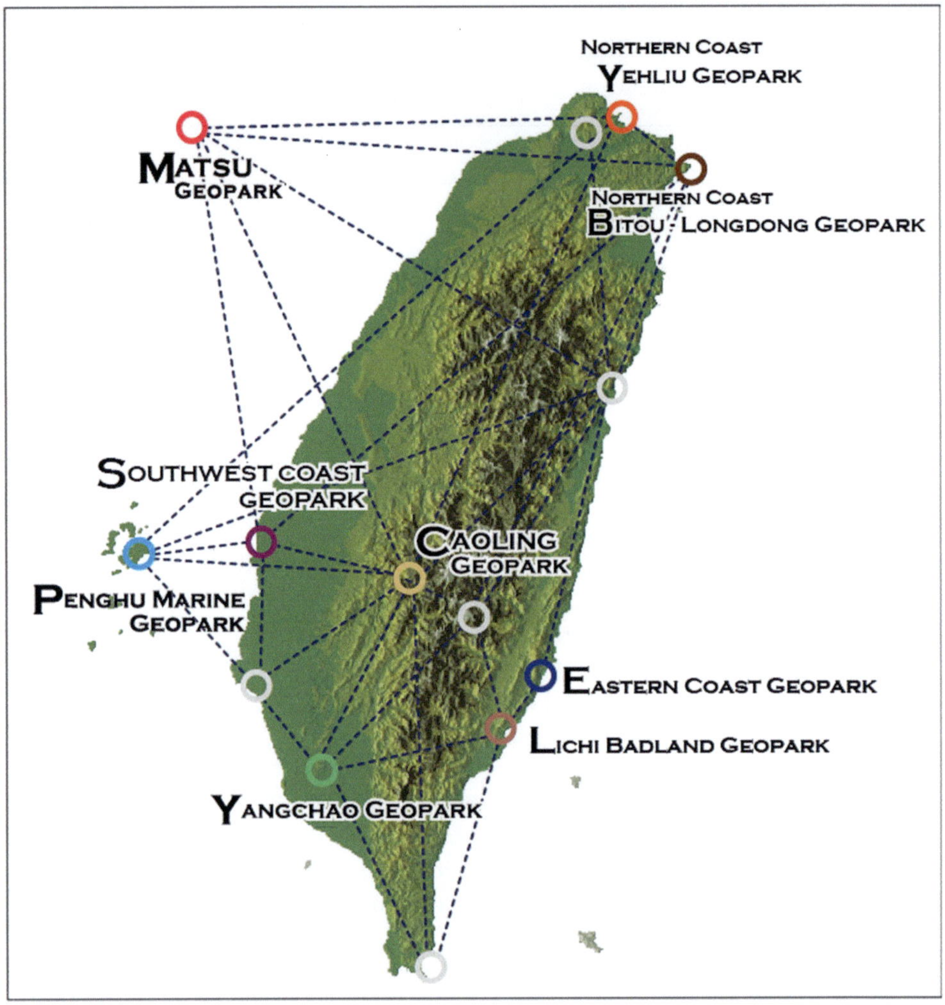

Fig. 1.1 Taiwan's nine geoparks as of 2018

Fig. 1.2 Group photo at the inaugural meeting of Taiwan Geoparks Association, March 2017

Table 1.1 The main characteristics of the Taiwan's geoparks

NO.	Name of Geopark	Major Landscapes	Geological nature	Major geomorphological Characters or values	Geo-Products
1	Northern Coast Yehliu Geopark	Coastal landscape	sandstone	geodiversity with weathering and erosion landscape	crab, sea food, and fish products
2	Northern Coast Bitou-Longdong Geopark	coastal landscape	quartzite and sandstone	marine hot spring, geo-diversity, marine terrace and platform	sea food and other marine products
3	Penghu Marine Geopark	coastal landscape	Basalt rocks	stone weirs, columnar joints	cactus, sea food, and fish products
4	Tsaoling Geopark	landslide landscape	sedimentary rocks	dip slope, waterfall, landslide	tea oil, bamboos, coffee, and other agricultural products
5	Yanchao Mud Volcano and Badland Geopark	Badland	mudstone	mud volcanoes and mud domes	guavas, promelos and Jujubis
6	Lichi Badland Geopark	Badland	Lichi Mélange	tectonic collisions, rills and gullies, badlands	guavas, and sugar apples
7	Matsu Geopark	coastal landscape	granite, diorite	sea arches, sea stacks, architecture, military,	liquor and sea food
8	Southwestern Coast Geopark	coastal landscape	Alluvial rocks	sand dunes, sandbars, lagoons, wetland, salt fields	salt products and sea food
9	Eastern Coast Geopark	coastal landscape	sandstones	coral, rocks, fishes	tuna, flying fish, and indigenous crafts

Fig. 1.3 Basaltic columns at Tongpan islet

Taiwan on its way for sustainable development with both horizontal and vertical integration of governance.

The proposal to establish geoparks came to Taiwan scholars in 2004 when Professor Shin Wang of National Taiwan University, led an academic team to the first Conference on Global Geoparks in Beijing, China. In 2005, under the support of Penghu County Government, the Geography Department of NTU held a World Heritage and Geopark Conference. Scholars and representatives from around the world and within the country agreed that the landscapes of Penghu are worthy of being developed into a geopark. Soon a commission for establishing a geopark was

Fig. 1.4 Basaltic columns

Fig. 1.5 Basaltic sea arch

formed in Penghu County. With the advancement of Penghu County Government's initiative, the Taiwan Geoparks Network (TGN) was advocated later based on the national inventory of special geological sites.

The realization of the Taiwan Geoparks Network was made possible at the National Landscape Conservation Conference in 2011, witnessed and supported by the Director of the Forestry Bureau and the Director of the Tourism Bureau. Six geoparks were identified in the conference, including the Penghu Marine Geopark, the Northern Coast Yehliu Geopark, the Tsaoling Geopark, the Yenchao Geopark, the Lichi Badland Geopark, and the Matsu Geopark. The Taipei Declaration for the promotion of geoparks was signed and came into effect in the same year. As of today, Taiwan has nine geoparks to promote its diversity of landscapes and to enhance its environmental quality (Fig. 1.1).

Each Taiwan geopark has its distinguishing features with scientific, socio-cultural and environmental values. Their landscapes all have close connection with the local ways of life and even people's livelihood. Penghu Marine Geopark

Fig. 1.6 Basaltic coast

features are basaltic landscapes and marine ecology, which formed the foundation of fishing with traditionally smart ways of utilizing the local geological materials. Tsaoling Geopark features are landslides and structural landscapes, which condition economic life of the locals (Lee 2014). Mud volcanoes characterize Yenchao Geopark and badlands, which most locals believe, are the reasons for their guavas to be tasty. Lichi Badland Geopark features are mudstone badland, which is boasted to be one of the only two mélange landscapes in the world for scientific reasons. Northern Coast Yehliu Geopark features are erosional coastlines and rocky landscapes, which most metropolitan residents enjoy for the weekends. The Matsu Geopark is famous for its granite and cold war heritage and socio-cultural landscapes. All of these landscapes are rare, unique and important for environmental education and scientific research, in addition to their significance to the local life. The landscapes are fundamental assets for local economies, cultures and even eco-systems.

The Taiwan Geoparks Network holds network meetings twice a year to promote the geopark concept and updated ideas among the geopark communities. During some of the geopark network conferences, field trips to Japan, Hong Kong and other geoparks were conducted. Such network meetings forced the motivation for the locals to reconsider the relationship of society and nature, and reached most consensuses for the adoption of geopark concepts as a tool for local sustainable development. The communities of Taiwan's geoparks benefited from learning with each other through networking activities. Knowledge for landscapes and site management strategies are shared and leads to a prolific and sound Taiwan Geoparks Network. Geoparks networks became a synonym term of the co-learning and learning community in the geopark society of Taiwan.

As of today, 9 geoparks are cooperating in the Taiwan Geoparks Network. In addition to the six mentioned above, there are the Southwestern Coast Wetland Geopark (or Yun-Chia-Nan Coastal Wetland Geopark), the Northeastern Coast Long-dong and Bitou Cap Geopark, and the Eastern Coast Geopark. With a strong local initiative and supportive academic endeavor, geoparks were formed with some encouragement from relevant central agencies. The procedures, however, for a geopark to pass the application and evaluation under the Cultural Heritage Act are strengthened due to the amendment of the Act in 2016.

The Taiwan Geoparks Association was established in 2017 in order to improve the coordination and determined development of geoparks (Fig. 1.2). Taiwan is not part of the global UNESCO framework because it cannot be a member of the UN due to its political relationship to China, but it tries to exactly mirror UNESCO's criteria and procedures related to Geoparks in order to fulfill its wish to look after its geoheritage sites on behalf of all people on the planet. These activities are seen as a gift to the world and as a responsible stewardship despite actual political difficulties.

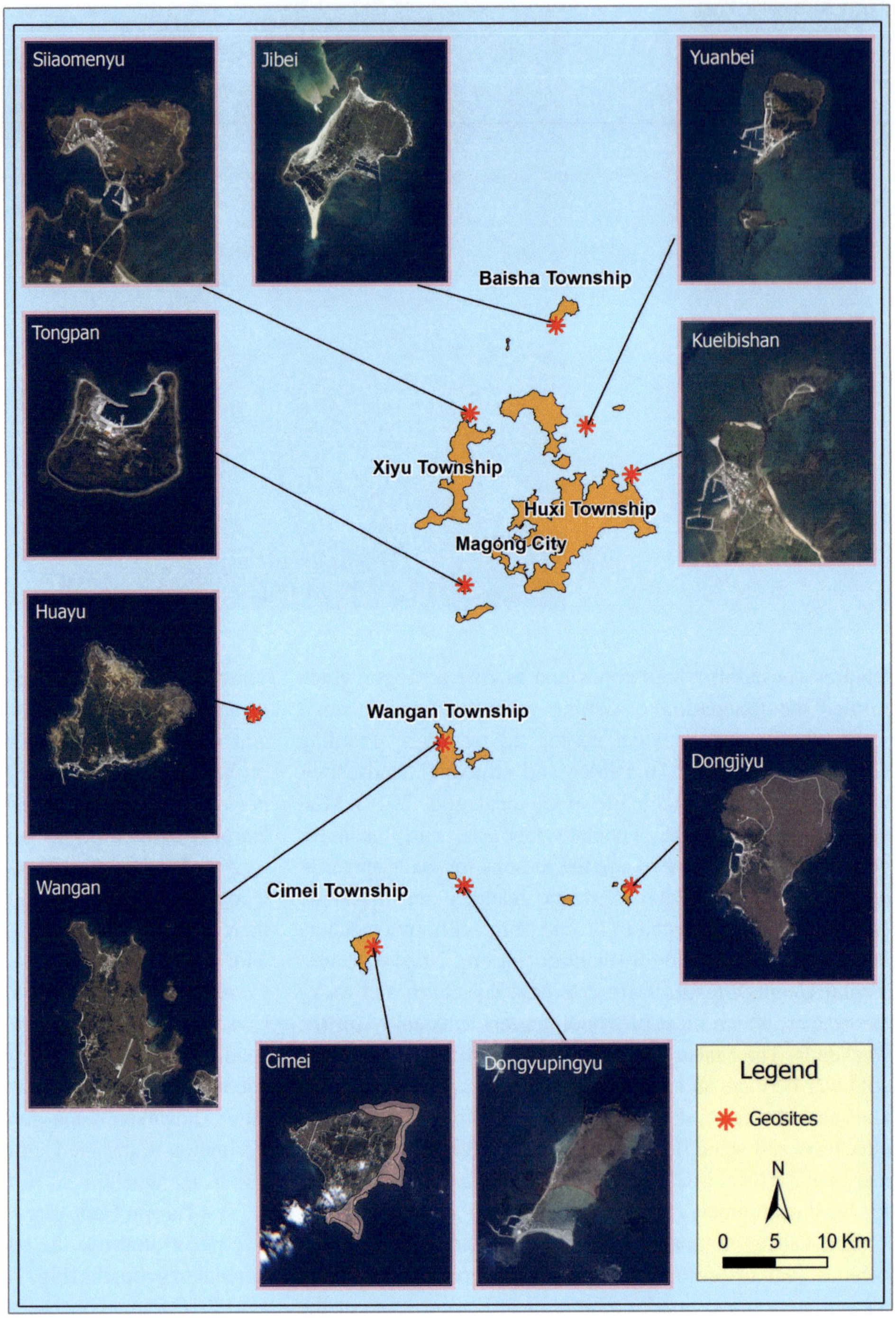

Fig. 1.7 Landscapes of the Penghu Marine Geopark

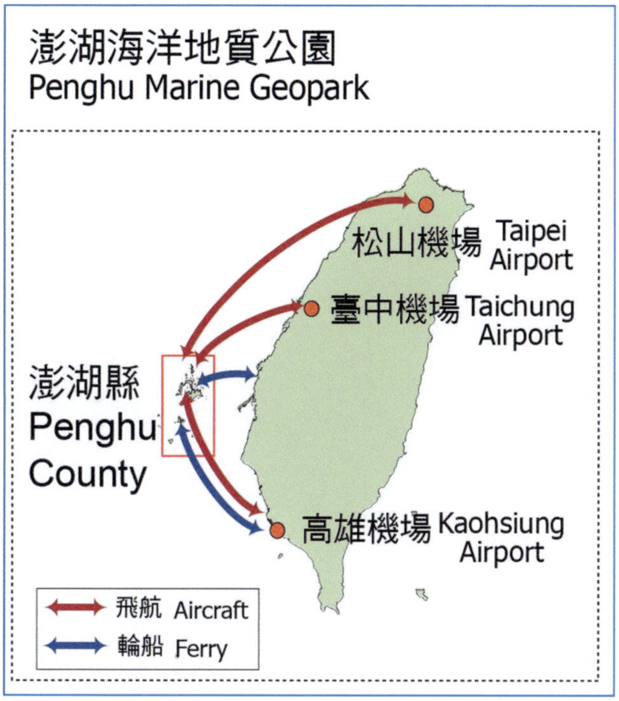

Fig. 1.8 Location map of Penghu Marine Geopark

1.2 Characteristics of Taiwan's Geoparks

There are different characters among the nine geoparks of Taiwan (Table 1.1). A brief knowledge of them all will ensure better understanding of how significant geoparks are in Taiwan's context and in its contemporary scientific relevance to the global society.

1.2.1 Penghu Marine Geopark

Penghu Marine Geopark is famous for its basaltic landscapes (Figs. 1.3, 1.4, 1.5 and 1.6) and marine ecology. Magnificent volcanic landforms ubiquitously exist in the archipelago of Penghu, such as hexagonal or pentagonal basaltic rock pillars and lava platforms. Unique cultural landscapes and economic activities nurtured by its particular geology, geomorphology, marine ecology and geography are characteristic and charming. The entire geological and geographical milieu gives Penghu Marine Geopark special socio-economic meaning and values (Liu 1999).

The Penghu archipelago is located in the Taiwan Strait (Figs. 1.7 and 1.8). It is mainly made of erupted basic basaltic rocks, resulting from gradually cooling magma. With the combination of special geological features, marine ecology, and human activities nurtured by the maritime history, the uniqueness and importance of the geopark is outstanding. The basaltic columns are different in shape, color and size all around the archipelago.

Among all geo-sites, the basalt is the main component. The basalt erupted between 18 and 8 million years ago with several centers. Paleo-soils are common in Penghu because between two eruptions there was enough time for the basalt to be weathered into soil (Fig. 1.9). After the soil was formed, another volcanic eruption occurred again. The

Fig. 1.9 Palaeo-soils evolved between two eruptions at Penghu

Fig. 1.10 Submerged Double-heart stone weir and its islet at high tide

Fig. 1.11 Double-heart stone weir for fishing at Chimei islet

junction between the hot lava and the cold soils produced spectacular chilled margins.

In early times, fishing was the main base economy in Penghu. Besides coastal inshore and offshore fishing, many families became involved in intertidal stone weir fishing. Therefore, the building skills of using interwoven coral stones and basaltic rocks for the fishing weirs of various shapes and sizes to suite the local environments become a great value and traditional wisdom (Su 2014). The skills for building stone fishing weir architecture are now part of the intangible heritage. The famous Double-Heart Stone Weir in Chimei islet, Couple Stone Weir in Erkan village, and the high density of various fishing weirs in Jibei islet are all good examples. They demonstrate great traditional wisdom accumulated through generations at the local sites in making good use of local materials (Figs. 1.10 and 1.11).

Fig. 1.12 A close-up look of coral reef wall as a windbreaker

Fig. 1.13 Coral reef walls as windbreakers for growing vegetables

Another example of traditional wisdom, as a modern attraction, is the characteristic style of local buildings. Typical Penghu architectures were mostly built with local basaltic rocks and coral stones. The combination of basaltic rocks and coral stones ensure that they lock one another to become stronger to defend against the strong wind of the archipelago. It demonstrates a wisdom gained from the severe living environment of the archipelago. The local techniques and aesthetics of building have become a feature worth exploring in geo-tourism.

Fig. 1.14 Exhibits at the Penghu Marine Geopark Center

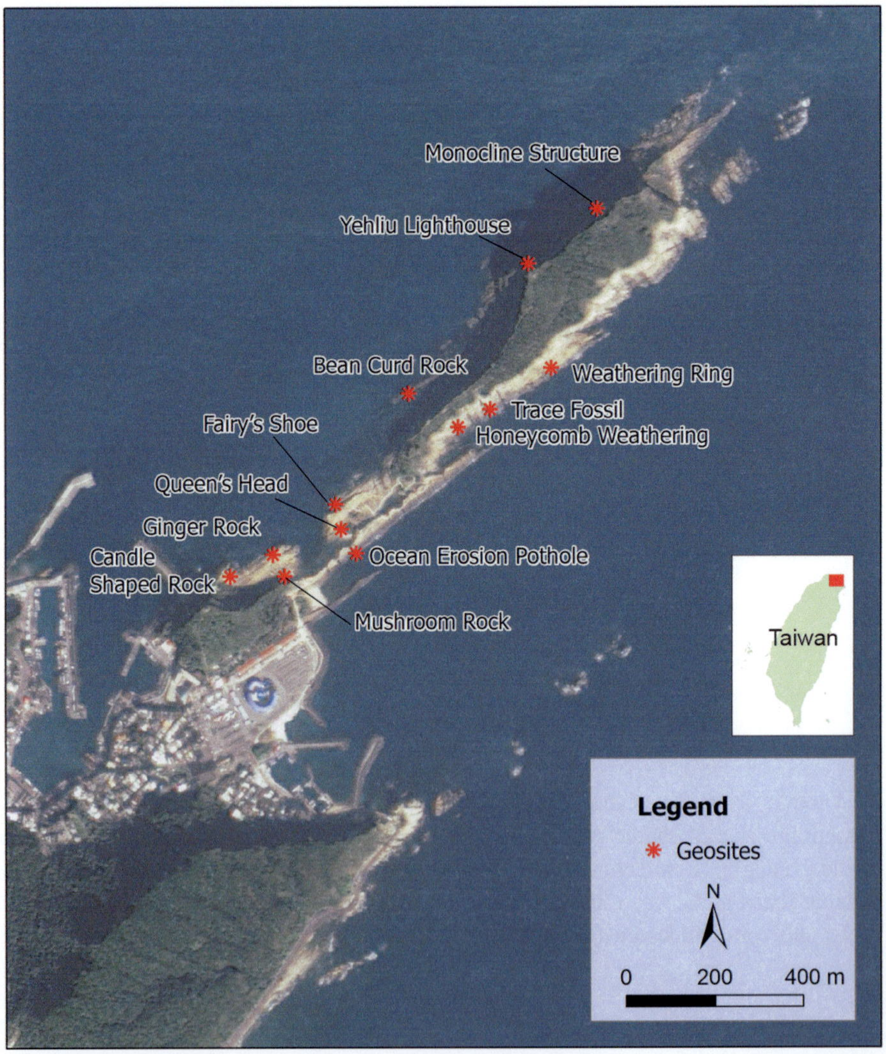

Fig. 1.15 Location map of Yehliu Geopark

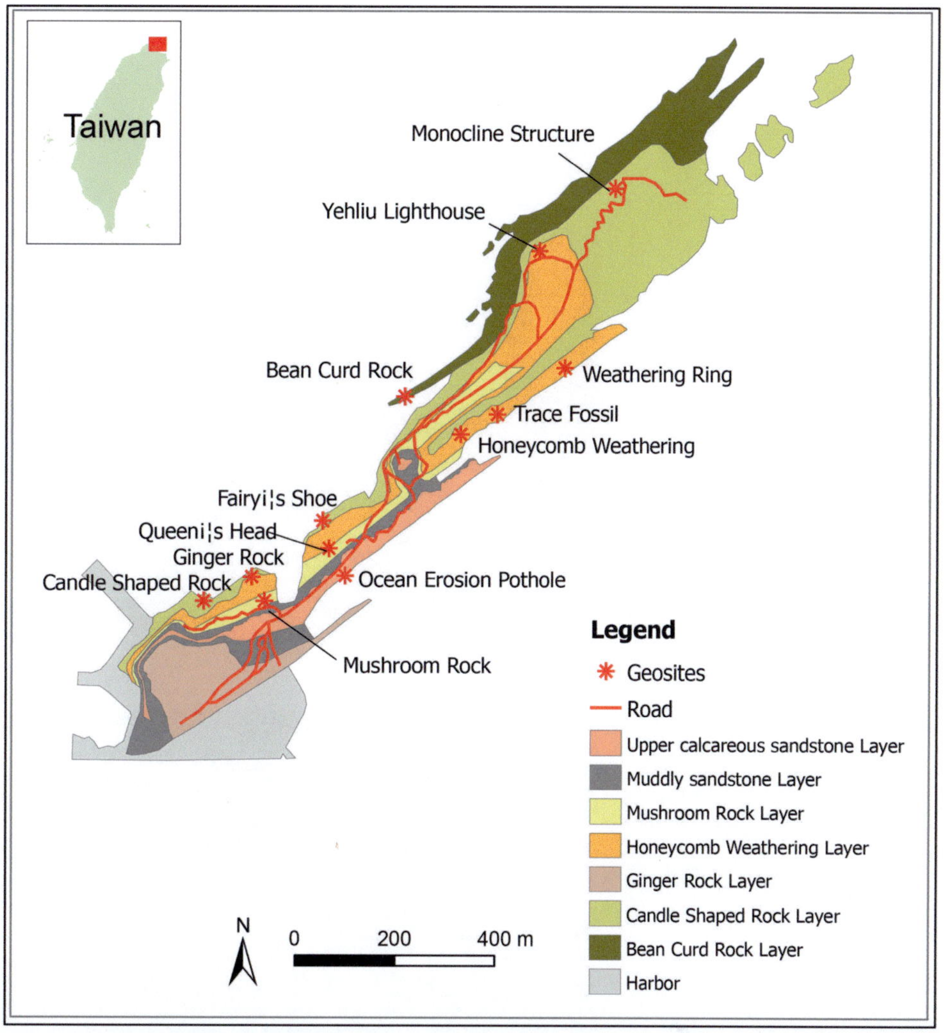

Fig. 1.16 Geological map of Yehliu Geopark

Fig. 1.17 Yehliu cape

Fig. 1.18 Dip slope at Yehliu geopark

Fig. 1.19 Marine pot holes

With particular strong winds in the winter, windbreakers are needed for essentially all agricultural and social activities in Penghu. Particularly important are the windbreakers against strong northern or northeastern winds for the growth of vegetables and for the comfort of settlement living (Figs. 1.12 and 1.13). Windbreakers made of coral and basaltic rocks are functionally significant because the breaker wall will become strong through time as the wind power will tighten the interlocking effects of the rocks.

Some local communities and residents show great concerns about Penghu's marine environment and landscapes, because it is obvious that its ecology and landscape is the foundation for fish and all sorts of seaweeds. Various local groups and associations are involved in local environmental affairs, such as the Nature Study Club, Wild Bird Federation, etc. These groups have been established to appreciate and to protect the local natural milieu. Penghu has been and still is an ideal habitat for migrant birds. Migrant birds, such as

Fig. 1.20 Honeycomb weathering textures

Fig. 1.21 Weathering rings

terns, visit Penghu annually for seasonal nesting and foraging. It makes bird watching activities highly cherished. As stakeholders, the bird watching groups are serious about nature conservation and environmental protection. An exhibit center of Penghu Marine Geopark is located at the center of the main settlement of the archipelago to foster environment education for landscape and nature conservation (Fig. 1.14).

1.2.2 Northern Coast Yehliu Geopark

Located in northern Taiwan, the Northern Coast Yehliu Geopark has many attractions, reminding people of a crown decorated with jewels. That is why the northern coast is called "the Crown Coast" (Fig. 1.15) decorated by a variety of natural and cultural landscapes (Fig. 1.16). Yehliu

Fig. 1.22 Sea urchin fossil

Fig. 1.23 Mushroom rocks

geo-site is a cape formed by strong sandstone and a headland stretching about 2 km into the sea located around a structure of cuesta slopes (Figs. 1.17 and 1.18).

The Yehliu Geopark is the first site in the Northern Coast transformed by following the UNESCO guidelines for geoparks characterized by grotesque-shaped rocks due to differential weathering and erosion. With the increase in tourism and its close proximity to the densely populated Taipei metropolitan area, endless streams of tourists flood in during public holidays and at weekends.

Yehliu Geopark is a park of natural wonders. Rocks carved by wave-cutting and weathering over the years were formed into shapes resembling real figures of people, plants or animals. The most famous one is named the "Queen's Head". Other "statues" are called the Fairy's Shoe, the Mushroom Rocks, the Tofu Rocks, and the Elephant Rocks,

Fig. 1.24 Candle rock caused by differential weathering

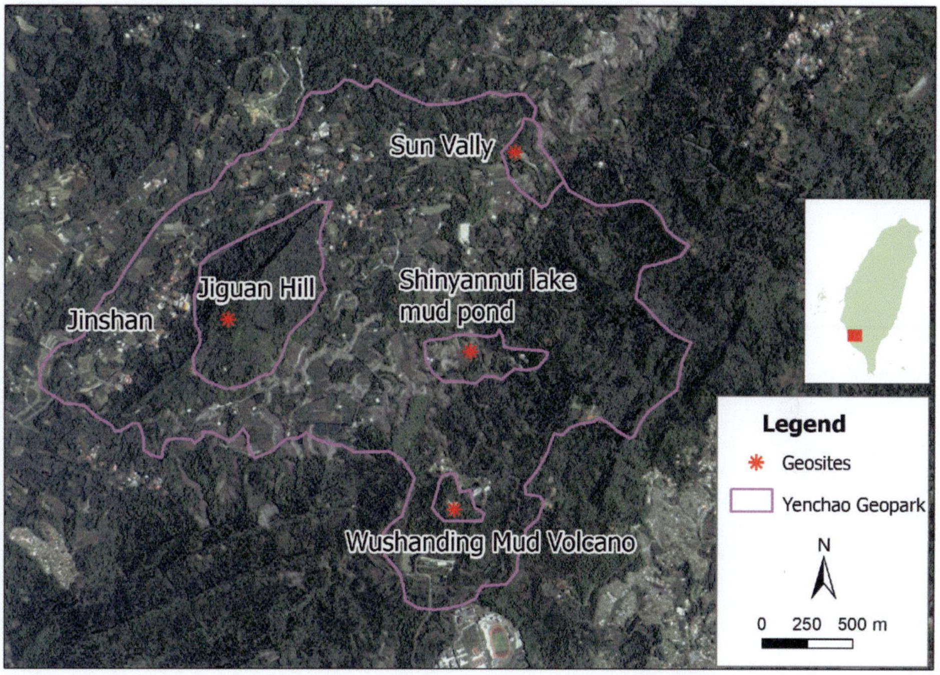

Fig. 1.25 Location map of Yenchao Geopark

along with many uniquely shaped potholes. Yehliu Geopark has many micro-landscapes. Rocks resembling ginger, chessboards, mushrooms, and candlesticks are just some examples. Fossils like echinoids or corals, as well as trace fossils are present in many corners of the park as well (Figs. 1.19, 1.20, 1.21, 1.22, 1.23 and 1.24).

1.2.3 Yenchao Mud Volcano Geopark

The Yenchao Mud Volcano Geopark, distinguished for its mudstone and badland, is situated in Kaohsiung City (Fig. 1.25). Major scenic spots include the Wushanding and Kuansuiping mud volcanoes and the Shinyannui lake mud

Fig. 1.26 Fresh mud flows just out of the mud volcano

Fig. 1.27 Mud volcano at its highest height

pond. In accordance with the Cultural Heritage Act, the Wushanding mud volcano was designated as a Natural Reserve Area as of March 2003. Mud volcanoes are pseudo-volcanoes where underground natural gas, liquid and mud are under pressure and erupted along earth fractures to form micro-landforms. Mudstones are made of microscopic silt- and clay-sized particles; they dry up to form small pictorial patterns and pile up to form cone-shaped "volcanoes", both are attractive to visitors. The co-existence of mudstone and mud volcanoes makes environmental education for children great fun. With close proximity to Kaohsiung metropolitan area, the Yenchao Mud Volcano Geopark is famous for its bad lands with fascinating people and delicious fruits.

Yenchao is famous for its rugged landscapes as well. Each year between October and April, due to the shortage of rainfall, the badland areas are dominated by gullies and little farming takes place (Figs. 1.26, 1.27 and 1.28). Local residents build low dikes and reservoirs to keep water for irrigation. Badland reservoirs and ponds are landscapes of

Fig. 1.28 Bubbling of mud volcano at Wushanding

Fig. 1.29 Water reservoirs for irrigation are constructed with local wisdom to avoid muddy water to get into the agricultural field

natural and cultural structures, showing how people adapt to the badland environment (Figs. 1.29 and 1.30). Some local farming reservoirs are constructed to keep only water from precipitation but avoid mineral rich runoffs coming into the reservoir with muddy clay directly. It is local traditional wisdom accumulated through time.

Mud Volcanoes are deposits created by the geo-excreted liquids and are composed of underground mudstone and water. The lifting force of liquids results from the tectonic stress and methane gases emitted from the well-developed fault zones and fold axis. The shape of mud volcanoes is determined by the viscosity (water content) of the mud. The major shapes of mud volcanoes include mud holes, mud cones, mud shields, and mud basins. Apart from the shapes of mud volcanoes, the emission activities of mud, mud flows, and subsequent shrinkage help produce unique and sometimes interesting features. In addition, earth fingers formed by rainfall erosion, nature bridges, and small gorges/canyons are geological features unique to mud volcanoes.

Agriculture and Tourism are now the two main economies of Yenchao. Agriculturally, this badland consists of

Fig. 1.30 Landuse at Mudstone area of Yenchao geopark

Fig. 1.31 Jujubis at Yenchao geopark

soils from mud volcanoes of slight gray quality, containing sodium (sulfates) and magnesium oxide suitable for guavas, jujubes and grapefruits. Guava trees take up 1600 ha and can be harvested all year round. Jujubes are a specialty of Yanchao in winter. Grapefruits, guavas, and sweet jujubes (dates) are regarded as the three treasures of Yenchao (Fig. 1.31).

As for tourism and environmental education, all major sites at Yenchao has convenient accessibility to enhance its potential. Landscape and environmental interpretation is a thriving activity as the local cultural association takes the lead in educating local community members to become capable interpreters. The combined development of tourism and environmental education is the center of the landscape economy, where the local interpreters develop excellent interpreting materials and monitor the area's changing landscape (Fig. 1.32).

1.2.4 Lichi Badland Geopark and Eastern Coast Geopark

The Lichi Badland Geopark and East Coastal Geopark are located very close to each other. However, they have very different features worthy of understanding in their own light (Fig. 1.33).

Fig. 1.32 The mud-volcano at Yenchao geopark is an asset for environmental education

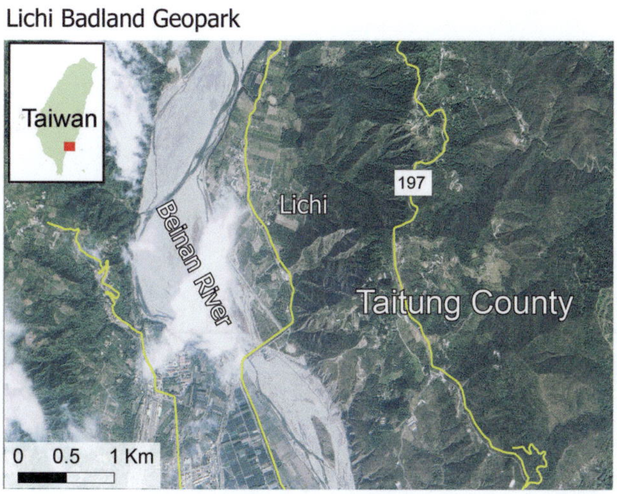

Fig. 1.33 Location map of Lichi Badland Geopark

The Lichi badland is not only significant for geological research for the subduction of convergent plates, but also valuable for eco-tourism. Earth scientists can observe the marks resulting from plate collisions and the rock formations of different oceanic crusts (Fig. 1.34). In the area of the Lichi badland, there are large amounts of developed gullies. The nature of the gully landscapes and their erosion is significant in research on the effects of convergent plates (Fig. 1.35).

The Lichi Badland Geopark delivers important proof of the collision of the Philippine Sea Plate and the Eurasian plate, in the character of the Lichi Mélange. It has long been significant for international earth science and geological research. The Lichi Mélange is widely distributed in a long stretch along the west side of the Coastal Range, extending from the north at Yuli in Hualien to the south at Beinan in Taitung County (Fig. 1.35).

The Lichi Mélange is composed of grey shale, mud rock or argillite, mixed with various foreign rocks. The foreign rocks of various sizes resulted from tectonic collisions, in which the sandstones (originally deposited on the edges of the continental plate) and ultrabasic rocks of the ophiolite suite are mixed. At the time of collision, the shearing function caused the basic muds to form scaly foliation. The directions of the shearing and the tectonic stress have proved identical, and the collisions are considered to have occurred in the Pleistocene epoch. Due to the lack of clear bedding in the argillites, the badland landforms can be easily eroded by flushing waters and thus commonly form gullies, earth fingers and natural bridges (Figs. 1.36, 1.37 and 1.38).

Due to that muddy badlands are not really suitable for planting and the fact that soil is poorly developed, the agricultural use of areas surrounding the geopark is difficult. Consequently, the cultural landscapes are mainly shaped by its geomorphological processes. Specific aboriginal cultures as well as unique badland agricultural practices have been thus developed in the area. Recently Lichi has nurtured many fruit orchards, with the policy support of farmers' association. Among the well-known and popular fruits are guavas and custard apples. Some fruits are labeled as geo-products (Figs. 1.39 and 1.40). The rhythm and tempo of community life become accordant to its natural environment and cultural/economic wisdom and make it attractive and sustainable.

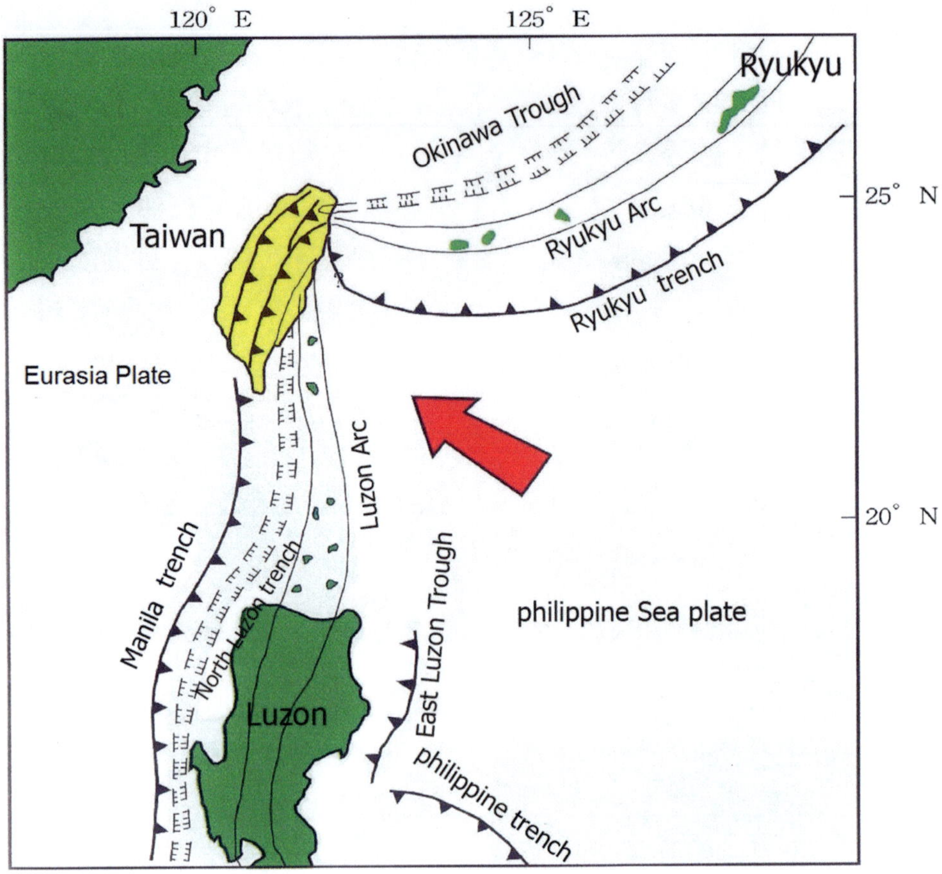

Fig. 1.34 Tectonic movements around Taiwan. The Lichi badland geopark is situated between Eurasia Plate and the Philippine Plate

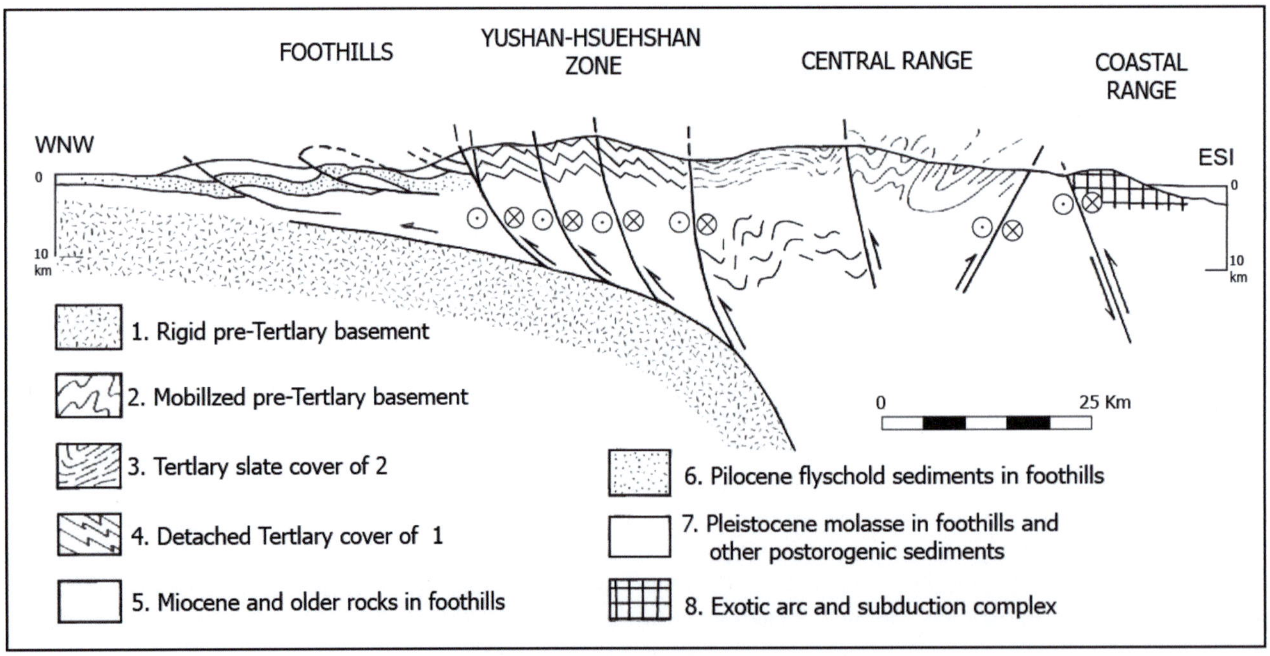

Fig. 1.35 The Eastern Coastal Range is the effect of convergent plates

Fig. 1.36 Lichi Mélange at Lichi Badland geopark

Fig. 1.37 Rill development at the Lichi Formation with the inputs of sandstone during crustal movements

1.2.5 The East Coast Geopark

The East Coast Geopark, located in coastal Taitung County, extends north to south. The combination of coastal mountain rocks and the layers have recorded the evolution of the island. This includes volcanic regions, rapid declines in the accumulation of thick sediment, and sediment generated by plate collision. It is also a place to observe scattered coral reefs around and on volcanic rocks and how they were slowly uplifted (Figs. 1.41 and 1.42).

"Shitiping" is a typical site to display the dynamic geological formation. It has a huge area of coastal terraces composed of volcanic ash and tuff. Its landscapes include the uplifted terrains, wave-cut platforms, coral reefs, sea

Fig. 1.38 An exotic block (the yellow part in the center of the picture) in mudstone

Fig. 1.39 Longan fruit at Lichi Badland Geopark

Fig. 1.40 Sugar Apple fruit is a typical cash crop at Lichi Badland Geopark

Fig. 1.41 Location map of Eastern Coast Geopark

trenches, caves and cliffs. The potholes, in particular, are one of Taiwan's Number one geological sites (Figs. 1.43, 1.44 and 1.45).

Bahsien Cave

Bahsien Cave is both of geological and archaeological interest. It is located on the cliffs that face the Pacific Ocean in Shanjian Village in Changbin Township. There are dozens of natural sea caves originally generated by the sea surf. Now they are scattered around 110-m-high cliffs with a rich Paleolithic pottery culture, named the Changbin Culture. It is the oldest prehistorical site found in Taiwan (Figs. 1.46, 1.47, 1.48 and 1.49).

Sanhsientai

Sanhsientai is located in the north of Chengong Township in Taitung County composed of offshore islands and coral reefs. Sanhsientai is an agglomerate rock that now forms a headland (Fig. 1.48). Gradual erosion has cut the neck that connected the mainland to the main island, which became a region of offshore islandsrich in biodiversity. Due to a lack of human disturbance, screw pine trees can be seen everywhere on the off-shore island.

Hsiaoyehliu

The name Hsiaoyeliu was gained because the topography and rock types of the area look similar to those of the Northern Coast Yehliu Geopark. Literary translation, Hsiaoyeliu mean "little Yehliu". However, a movement of local identity has made it a case for changing the name to mean something local. In terms of geomorphology, differential erosion of sandstones makes the place a good learning site for the young curious minds (Figs. 1.49, 1.50 and 1.51).

With the close proximity of the Lichi Badland Geopark and the East Coastal Geopark, the community members have good practice for knowledge exchange, thus learning from each other. For instance, at geopark networking meetings, they work closely together (Figs. 1.52, 1.53, 1.54 and 1.55). The working relationships of the two communities prove to

Fig. 1.42 Successive vertical sea notches demonstrate the uplifting movements of eastern Taiwan

Fig. 1.43 A close-up view of volcanic ash and tuff at the Shitiping site

be strong and robust as the ethnic combination and diversity of the two usually bring creative results in many occasions.

1.2.6 Tsaoling Geopark (Caoling Geopark)

Tsaoling Geopark, also called Caoling Geopark, is located in Gukeng, Yunlin County (Fig. 1.56). It is well-known for its natural beauty and the dynamic landform evolution of landslides, huge cliffs, waterfalls and potholes. Local governments subscribe to the conservation of the landscape and environmental education, and develop geo-tourism to boost its social and economic development (Lee 2011).

The infamous Tsaoling landslide of 1999 is situated on the southwestern dip slope of the Geopark (Fig. 1.57). This landslide is among the largest in Taiwan in terms of total area and magnitude. The landslide happened during the "921" Earthquake in 1999, covering an area of more than

Fig. 1.44 Volcanic ash and tuff at Shitiping makes stone staircases for visitors to enjoy the site

Fig. 1.45 Cross-bedding of volcanic ash and tuff at Shitiping site

4 km^2. This landslide is the fifth event that has occurred during the last 150 years. The high frequency of landslide demonstrates clearly the hazardous nature of the area. It is prone to earthquakes and persistent rains (Lee 2013). Surprisingly these landslides have been a critical education and conservation opportunity for the local community (Chen 2017). They reinforced the will for constructing a geopark by using the characteristic values of a geopark as important tool to educate the people at large on the significance of and threats through natural hazards in Taiwan (Lee 2014).

Bamboo groves are also a major feature in the Geopark. Bamboos are located ubiquitously, and the air is infused with their fragrance. Many local residents skillfully make bamboo crafts (Fig. 1.58), wood and stone carvings. Crafts, such as bamboo containers, bamboo ladders, bamboo column beads, stone mortar, stone tables, stone benches, and stone walls demonstrate traditional handicraft techniques. In addition, dried bamboo shoots and jelly figs are Tsaoling's famous specialties. These geo-products are big selling points of the Geopark.

Fig. 1.46 Marine sea cave at Bahsien Caves site

Fig. 1.47 Sea cave at Bahsien cave site

1.2.7 Matsu Geopark

Matsu islets, located to the north-west of Taiwan and off shore of China, faces China's Min River estuary. Matsu' s five major islets and numerous inhabited and uninhabited islets with ecological significance are divided into four townships. The islets were geographically separated from the continent of China around 8000 years ago at the end of the ice age when sea level rose about 120 m around this area.

Fig. 1.48 Sanhsientai islet connected by a foot bridge

Fig. 1.49 Coastal landscape at Hsiaoyehliu

The Matsu archipelago is the northernmost territory under Taiwan's jurisdiction. The Beigu islet off shore Luohan Flats on northern Xiyin islet (at 26° 22′ 58.8″N and 120° 28′ 34.0″ E) is known as the "Northernmost Frontier of Taiwan". When the weather is clear, the Beigu islet may be seen from the scenic outlook of northern Xiyin islet (Fig. 1.59).

The abundant resources of Matsu include marine, ecological, and cultural resources and war relics. Fishing was the main economy in historical times. Now migrant birds are the big attraction to both the locals and tourists. Terns of various sorts, in particular the Chinese Crested Tern attract seasonal tourists (Fig. 1.60). Plants like the Matsu *Lycoris radiates* are unique and not commonly found elsewhere. The Mingdong architectural culture and military structures record the history of the last century (Fig. 1.61). The military underground tunnels have become a characteristic part of 'black' tourism. As war relics they are not only of local importance, but also of global significance as they symbolize the Cold War (Lin 2013).

Matsu is mainly made of granite (Fig. 1.62), with the exception of Dongying mainly composed of diorite and Xiju composed of rhyolite. These plutonic rocks manifest a body of intrusive igneous rock with various SiO_2 contents. These rocks have been through erosion and weathering, and display great diversity of landscape, including magnificent crags and cliffs, sea trenches, sea arches, sea stacks and

Fig. 1.50 Differential erosion on sandstone at Hsiaoyehliu

Fig. 1.51 Erosional textures of Hsiaoyehliu

caves. All reflect the power of the marine processes, which control this coast (Fig. 1.63).

Matsu has a critical strategic position and lengthy war history. It was closed to civilians for decades until 1992 when the Military administration was dismissed and in 1994 it was reopened for tourism. Of all the battlefield landscapes, the most amazing is its military tunnel systems. Matsu has a high density of tunnels. Beside the accessible North Sea (Beihai), Wusha and Andong Tunnels, there are many tunnels and military garrisons unrevealed, and unknown to the public, which hide the blood and tears of the soldiers.

Tunnels are the hidden and significant landscapes that demonstrate to the young generations the past military conflict and confrontations. Many tunnels, famous for

Fig. 1.52 Field trip of a geopark network meeting at Hsiaoyehliu

Fig. 1.53 A guided tour at Eastern Coast Geopark

tourism, serve as great geo-tourism and geo-conservation examples. A few of them are illustrated here.

Tunnel 88

Opening the gate of Tunnel 88 is like opening the door to a secret treasure trove. Legend has it that this tunnel was where the local people once took refuge from pirates. During the military administration era, the Taiwan government tried to turn it into a military tunnel without success. When it was finished in 1974, after 10 years of work, it was named Tunnel 88 to commemorate a former Taiwan President Chiang Kai-shek's 88th birthday. The tunnel naturally maintains a stable temperature of 15–20 °C, making it perfect for the storage of old wines (Figs. 1.64 and 1.65).

Wusha Tunnel

The deep, dark Wusha Tunnel, or Wusha Beihai Tunnel, was built in 1968, at the same time as other Beihai Tunnels on

Fig. 1.54 Lichi community members pounding rice for making rice cake during a network meeting in 2014

Fig. 1.55 Lichi community enjoying foods after a network meeting in 2014

Nangan and Dongyin were built. It is 550 meters long and 9 to 15 meters wide. While it is not as large as the Nangan tunnel, it was dug laboriously out of the solid granite by soldiers using only pickaxes, shovels, iron rakes, baskets, and other simple tools. The work went on for almost three years and cost the lives of over 100 soldiers. Canoeing in the tunnel has gained popularity in recent years, offering travelers the opportunity of an unforgettable experience (Fig. 1.66).

Beihai Tunnel

Nangan's Beihai (North Sea) Tunnel runs from the Tieban coast deep into the heart of the hills in a mesh shape. The tunnel's water-channel section is 10 m wide and 640 m long, and 18 m from the bottom to the top of the tunnel. The depth of the water is eight meters at high tide and four meters at low tide. The tunnel path is 700 m long and takes about 30 min to walk around. Originally designed as storage space

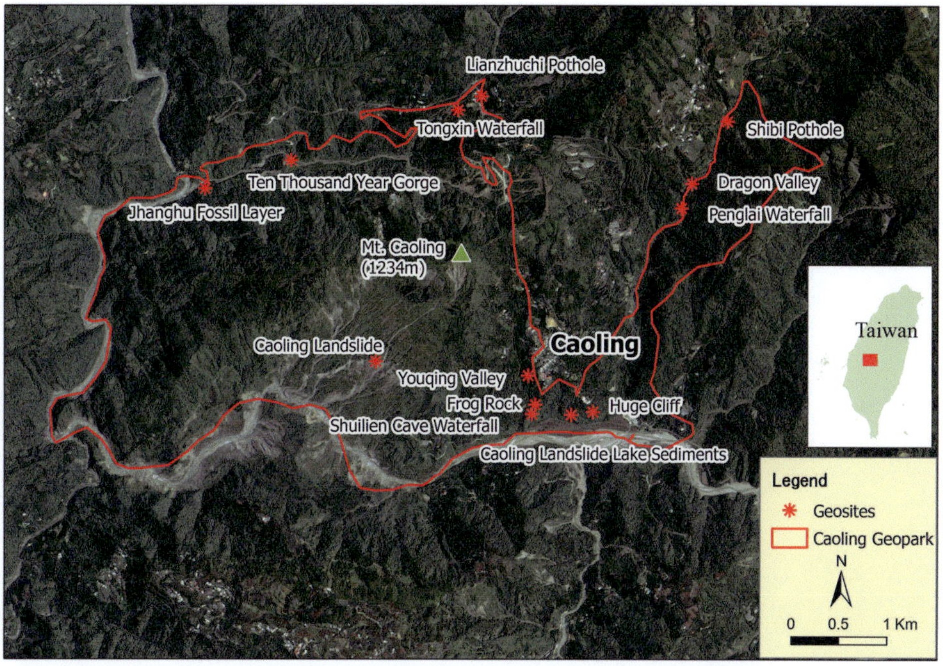

Fig. 1.56 Location map of Tsaoling Geopark

Fig. 1.57 Dip slope at Tsaoling Geopark

of battle crafts, it is big enough to accommodate more than 100 small landing vessels. It took 820 days for two divisions, three infantry battalions, one combat engineer battalion, and one dump truck company were mobilized to work day and night for its completion. With the exception of a small amount of explosives, all of the work was done manually. Tunneling through the granite stone was arduous and dangerous. Quite a number of people lost their lives in the tunneling (Fig. 1.67).

Lighthouses are critical for islands and fishermen. Two historically significant lighthouses of Matsu have become imperative for geopark interpretation. Both the architecture

Fig. 1.58 Bamboo forest at Tsaoling Geopark

and the functions of the lighthouses are attractive to modern day tourists, although their historical and political significance was even more important for the local islanders.

Dongquan Lighthouse

Dongquan Lighthouse is also called Dongju Lighthouse. It was built in 1872 by British to guide trading ships to Fuzhou City when China was forced to open it and four other treaty ports to trade in the late Qing Dynasty. The body of the lighthouse was made of granite; it rises 19.5 m and casts a beam of light that can be seen 16.7 nautical miles (about 31 km) away. A unique feature of the lighthouse is a low; 30-m windbreak wall that connects it with its office structure so that workers going between the two can bend low behind the wall during times of high wind to keep their kerosene lanterns from blowing out. The Lighthouse Museum, opened in the English-style main building in June 2008, displays a rich collection of lighthouse artifacts (Fig. 1.68).

Dongyong Lighthouse

Dongyong Lighthouse stands on the flank of Shiwei Hill of Dongyin islet. The locals call it "Dongyin villa". Completed in 1904, it is the northernmost historic registered building of Taiwan. The lighthouse and its accompanying main office building are in the 18th-century English architectural style. The surrounding milieu of the lighthouse has plenty of flora species to enjoy. It is considered the most romantic lighthouse building today (Fig. 1.69).

Architecture is often considered to be culture preserved through history. In Matsu, due to the military administration and the limited civilian ways of life, many architectural cultures are preserved and sometimes called the stolen time of Matsu civilian life. However today the buildings and structures of Matsu have become a symbol of architectural beauty.

The village of Qinbi is the best known village of Matsu islets and is also known as "a Mediterranean town on the Taiwan Straits". Another name for the village is Jingao, literally translated to "mirror harbor". The name derived from the crystal-clear waters of its bordering sea. Qinbi has the best-preserved eastern Fujian-style buildings in all of Matsu. It is very much appreciating by both the locals and foreign tourists and even architectural, cultural, and artistic circles. Chen family from the Fujian area originally developed it in the late Qing Dynasty. Granite rock steps, one orderly level above another, lead visitors up hill to the village. Strolling up the hill among the village buildings is a walk of history for tourists. To secure the roofs of the buildings from strong winds, most households used heavy boulders to top the shingles. With their shape resembling compact squares, each rooftop looks like a seal when seen from the top. That is why each Fujian-style stone house is named "a square seal". Rocky windbreakers are important in this place of windy winter, but more than often symbolic windbreakers of different animal shapes scatter in the village, and they are mostly made of granite stones.

Each of Matsu's islets has its own center with various settlements and traditional architectures. The difficulty of

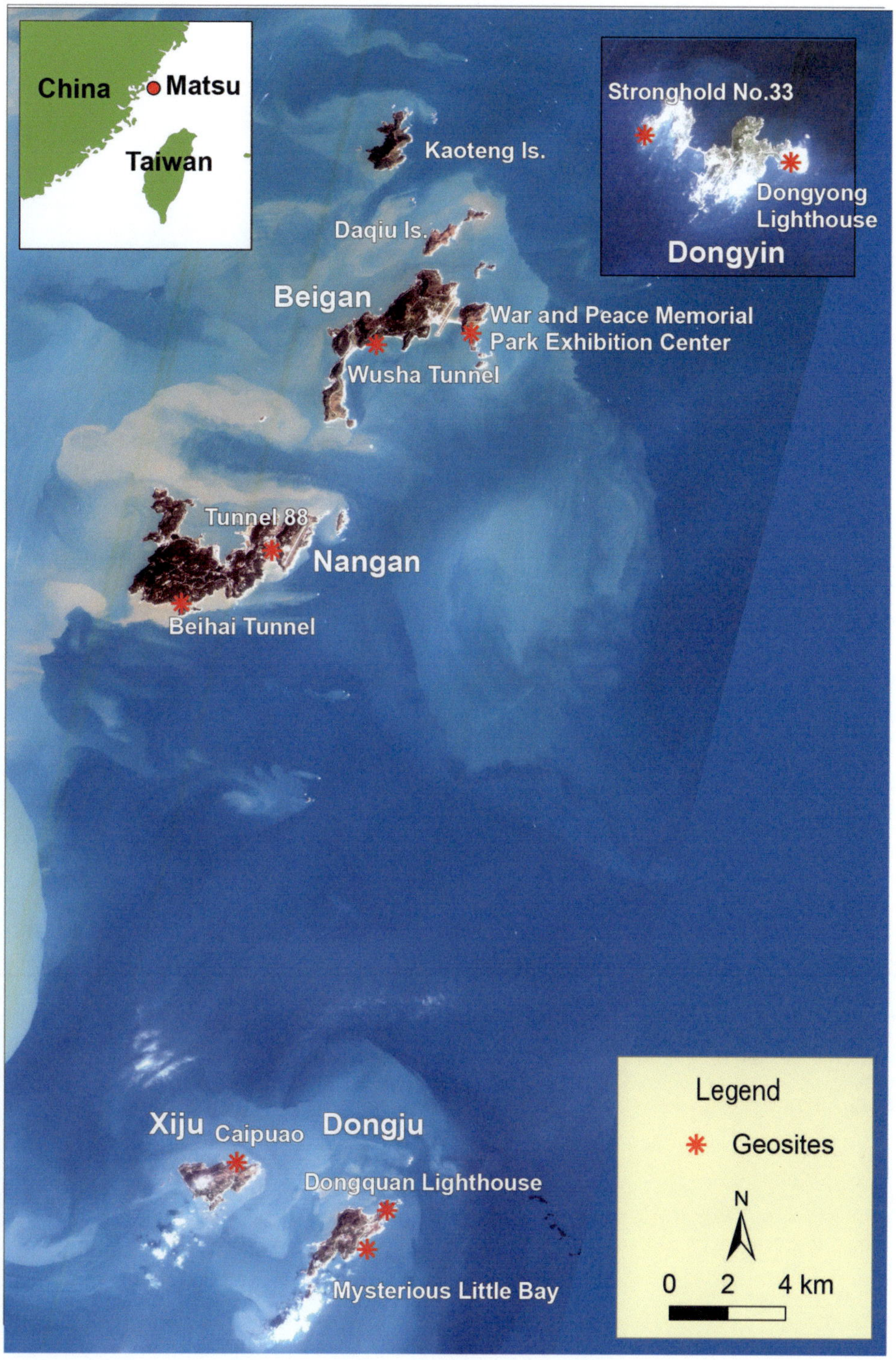

Fig. 1.59 Location map of Matsu Geopark

transportation between the islets contributed to the various building styles and settlement's spatial elements. Along with the military controlled era, most traditional buildings have been preserved in their original form. This is one reason why even today we may still spot well-conserved building and architectural styles all around the islets. The Fuzheng Village of Dongju islet is another significant case. Fuzheng village was once the most prosperous village of Dongju islet. However due to the gradual decrease in fish catches, residents of the village moved elsewhere. Today it has a small population but its stone houses still tell the ancient beauty of the fishing village (Figs. 1.70 and 1.71).

Military strongholds presented ubiquitously on the small Matsu islets. As they are artifacts built during the Cold War

Fig. 1.60 The Chung islet of Matsu is full of terns during summer time

Fig. 1.61 Slogan, meaning "to fight for the last victory", carved on strong granite to show the military spirits of the Cold War

era, many consider them not only symbols of strength but also symbols of the sweat and blood of soldiers. The locals consider them as a media for the pathway to the past because the strongholds showcase the difficult life and civilian military training of the past. Today they become attractions for tourism both for their geological foundations and military cultures.

Iron Fort, in Chinese language pronounced as "Taebao", might have been the most visited stronghold of Matsu due to

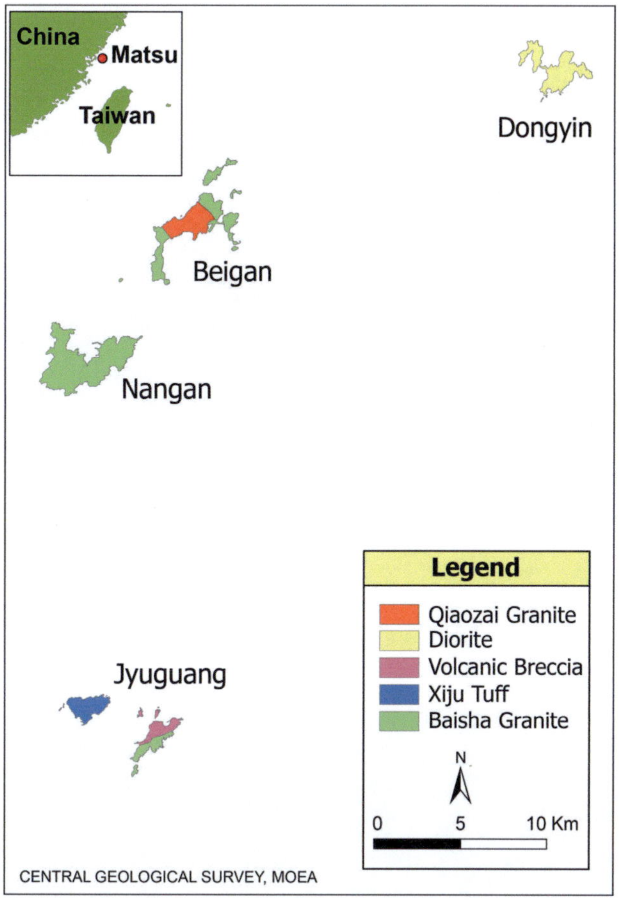

Fig. 1.62 Geological map of Matsu Geopark

Fig. 1.63 A scene of erosional coast in Dongyin islet

Fig. 1.64 Tunnel 88 at Nangan islet

Fig. 1.65 Rice wine bottles at the gate of Tunnel 88

its location. It is located on Nangan islet, along the coastline to the west of Renai Village. The Iron Fort was garrisoned for many years by elite amphibious frogman units during the Cold War era. Inside the stronghold tunnels are stone rooms, gun ports, cannon batteries, a toilet, and a kitchen, among other facilities, all packed into this small sea-defense fortification.

The Dahan Stronghold is located on the Taebao Coast of southern Nangan. Together with Juguang islet and other strongholds, it helped control the sea lanes. Construction of the stronghold began in 1975, with soldiers working in day and night shifts, using a small amount of explosives but mostly their own sweat and muscle to dig. It was finished in 1976. It has three levels including the top level for company

Fig. 1.66 Wusha Tunnel

Fig. 1.67 Andong tunnel at Dongying islet

headquarters, the second level for living space and machine guns, and the lower level designated for four 90 mm anti-aircraft guns, a briefing room, and storerooms. The tunnel is about 1.2–2 m wide, 2 m high for each level, and 430 m long (Figs. 1.72 and 1.73).

War and Peace Memorial Park Exhibition Center

The War and Peace Memorial Park Exhibition Center officially opened in 2010. It is located in Stronghold 12 on Dawo Mountain in Beigan. On a good day, the coast of China is visible from here. The War and Peace Memorial Park is Taiwan's first "war and peace" and "emotions and memories" theme park. One can wander through the old strongholds to view the surviving exhibits and historical materials, and discover the history of this frontier defense system.

Because of its geographic proximity to continental China, the Matsu islets have both island and continental characteristics, which have nurtured many distinctive species of

Fig. 1.68 Dongquan Light house with a long narrow wall as windbreaker

flora and fauna, such as the Matsu *Lycoris*, the Matsu Purple Sandalwood, the Matsu Wild Lily, and many other species named for Matsu (Fig. 1.74).

After the lifting of martial law and the withdrawal of the military administration, significant changes to the economic and social life occurred. Matsu became a brand new tourist attraction with impressive natural landscape, distinctive building styles, previous war landscape and tunnel systems. The island population used to be isolated from the rest of the world and the simple characteristics of the locals and original lifestyle add extra charm to the landscape. Today as tourists come for the Matsu Geopark, more tourist infrastructures or structures are added. Among them is the statue of the Goddess Matsu in Nangan. The statue shows Goddess Matsu

Fig. 1.69 Dongyong Lighthouse built by a British engineer in 1904

Fig. 1.70 Traditional houses with seal-like tops at Jinsha village

carrying a lantern standing on a wooden boat to represent her role in guarding and rescuing fishermen lost in the sea. The Statue is built in October 2009 with a height of 28.8 m, making it the world's tallest Matsu statue (Fig. 1.75).

1.2.8 Southwest Coastal Geopark

This wetland geopark is located in an area to the south of Budai of Chiayi County (Fig. 1.76). It is a bird sanctuary where migrant and seasonal birds along the western Pacific Ocean visit annually. It is a feeding ground for various protected migrant creatures. Nature lovers and bird watchers regularly visit the area. Many Taiwanese endangered species use the area for hatching and nesting. This area is also home to various small and returning fish and krill that is part of the base economy.

In the past, wetlands were considered as useless or marginal lands. However, with the concern of ecology and sound environment, we have come to realize that these

Fig. 1.71 A View of Fuzheng port

Fig. 1.72 Iron Fort used to be a stronghold for the amphibian frogman units

swamps, mangrove forests, and lagoons are important sources of food for fish, shellfish, crustaceans, and birds because of their slow water flow and the large amount of organic nutrients (Chi 2002). These wetlands are, in fact, complex and complete ecological systems with high productivity, abundant natural resources, and a high biological diversity. They constitute a rich gene bank and a favored habitat for large amounts of wildlife.

The mangrove is a major form of wetland vegetation. At coastal wetlands, saltwater and fresh water come together, so mangroves have evolved unique ways to survive in this brackish-water environment. Their thick leaves enable them to expel salt while storing water. The mangroves propagate viviparously and they have a system of roots comprised of supporting roots, underground roots, and breathing roots (Figs. 1.77 and 1.78).

Fig. 1.73 Dahan Stronghold on Nangan was responsible for guarding surrounding islets

Fig. 1.74 Strong winds control vegetation growth on Matsu

Birds attract the greatest attention of visitors to wetlands. Among them, the black-faced spoonbills that fly in every year to winter at Cigu are the superstars of the Southwest Coast National Scenic Area. There are only about 1700 of them left in the world, and a survey by Taiwan's Wild Bird Society in January 2003 shows that more than 700 of them passed the winter in the Zengwen River estuary wetlands. As the natural environment of the wetland has been improved,

Fig. 1.75 Statue of the Goddess Matsu, a symbol that guards fishermen

more black-faced spoonbills visit this wetland annually. They attract countless bird watchers both from Taiwan and overseas. In addition, in spring and summer each year, flocks of little egrets, cattle egrets, and black-crowned night herons gather and build nests in the windbreaks and mangrove forests, creating awesome scenes of "egret forests." Great flocks of these birds return to nest at dusk. With their fluttering wings covering the skies, they form an unforgettable sight (Fig. 1.79).

Tideland crabs and mud skippers are small animals frequently seen on mud beaches. With their large numbers and approachability, along with their unique form and method of motion, they have become star objects in wetlands ecology observation-especially for children. Four types of crabs are abundant here. The males wave their huge single-side claws in their most characteristic action. Clusters of monk crabs and mud-skipper fish, with their unique jumping motion, provide yet unforgettable ecological sights.

The area had excellent salt fields since 1784, which contributed to its fame in literature and ways of life. Culturally and socially, it was part of historical rural affluence. Due to the competition from modern salt manufacturers, the fields were closed in 2001. More than 200 years of salt making history was preserved and turned into coastal nostalgia. The salt fields explain why the area is good for bird sanctuaries. Because the shallow salt fields along with the wetlands and swamps had a rich benthos, they provided migrant, seasonal and resident birds with safe nests and feeding. Some salt fields with deeper water became good places for harvesting small fish and krill (Fig. 1.80).

1.2.9 Northeast Coastal Longdong and Bitou Cape Geopark

The abundant geographical resources of the Longdong and Bitou Cape geopark (Fig. 1.81) offer surprises at every corner. In a way, the geopark is a miniature projection of Taiwan, as it developed from a sunken land on the coast of South China pushed up from the sea and turned into a mountainous island. The diverse coastal landforms along the geopark continue to reveal the unique nature of this area's geological background and its development. It is indeed a perfect place for appreciating the beauty of the land.

The origin of the name Longdong (literally meaning "dragon's cave") comes from the shape of the region's coastline, similar to the shape of the head of a lying dragon. Along the coastline are a series of hard, steep-rising sandstone formations. These rocks with clear-cut joints stand erect and close to the ocean, vigorously announcing their presence. The impressive sea cliffs are composed of extremely hard sedimentary rocks. They are characterized by their rich and diverse geological views and recreational water activities. The abundant erosion platforms here serve as a great natural classroom for observing shore life and various types of algae.

From Bitou cape and Longdong geosites to the Cape of San Diego forms one splendid coastline (Figs. 1.82 and 1.83), which is a typical "Ria" landform with rock formations sculptured by erosion, abrasion, weathering, displacement, and other natural forces. The ria landform showcases mushroom-shaped rocks, tofu-like chessboard rocks,

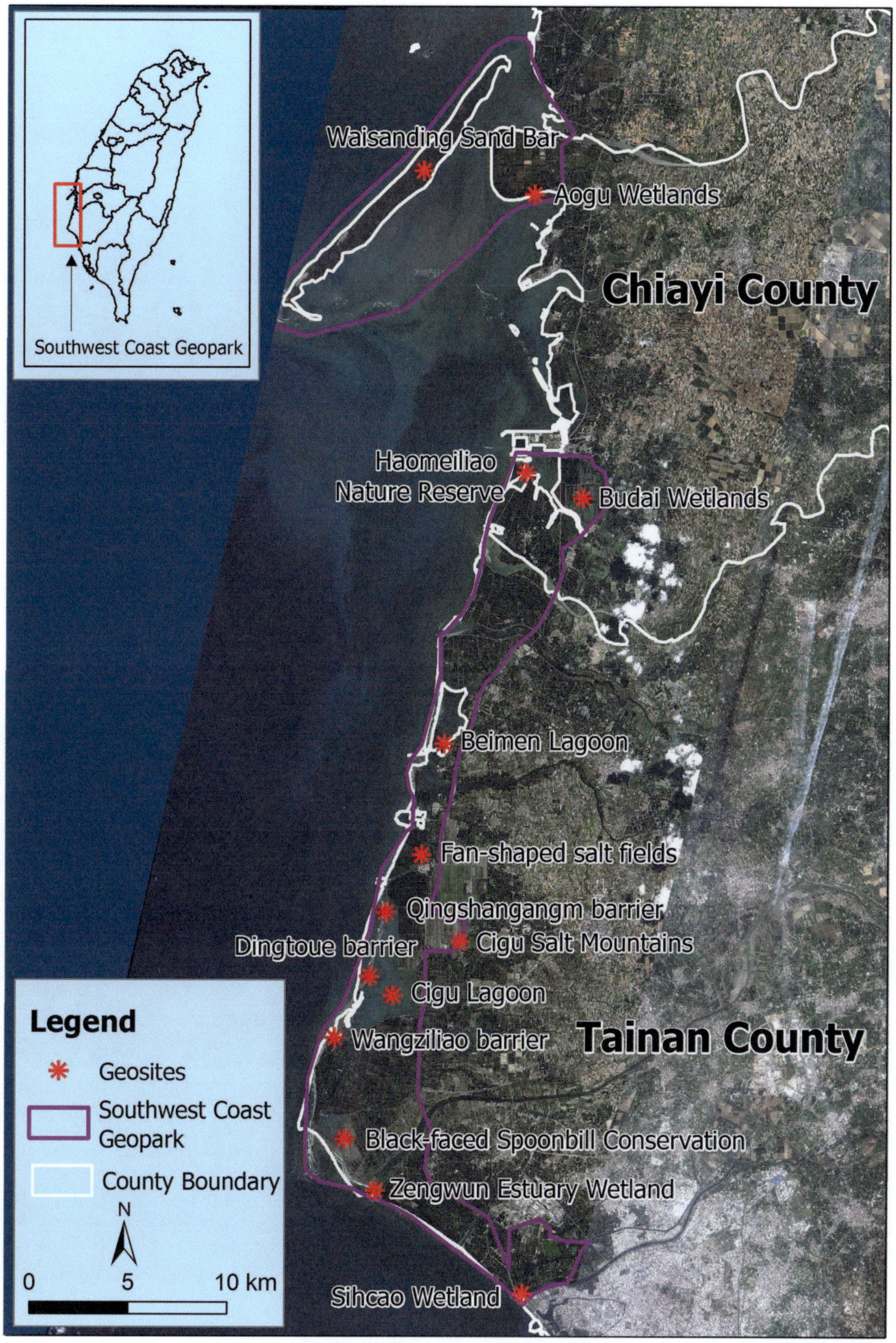

Fig. 1.76 Location map of Southwest Coastal Geopark

hive-like honeycomb rocks and other naturally formed rock formations.

For instance, the erosion-formed, striated rock formations at the Nanya area are typical attractions to many. Bitou Cape's intriguing mushroom rocks including sea cliffs, platforms, undercut bluffs, and other erosion landforms (Figs. 1.84, 1.85, 1.86, 1.87 and 1.88). Longdong Cape has typical steep sandstone cliffs above sea cut platforms (Figs. 1.89, 1.90 and 1.91) which both amateur and professional fishermen consider the perfect sites to enjoy fishing.

The fishing industry is an important economy and sampling local seafood is a must for the tourists in this geopark. A three-mile stretch of golden beach between Yanliao and

Fig. 1.77 Canal with mangroves along its bank in Southwest Coastal Geopark

Fig. 1.78 Mangrove in the Southwest Coastal Geopark

Fig. 1.79 An aqua-farm with many oyster racks at Southwest Coastal Geopark

Fig. 1.80 Scene of salt fields at Southwest Coastal Geopark

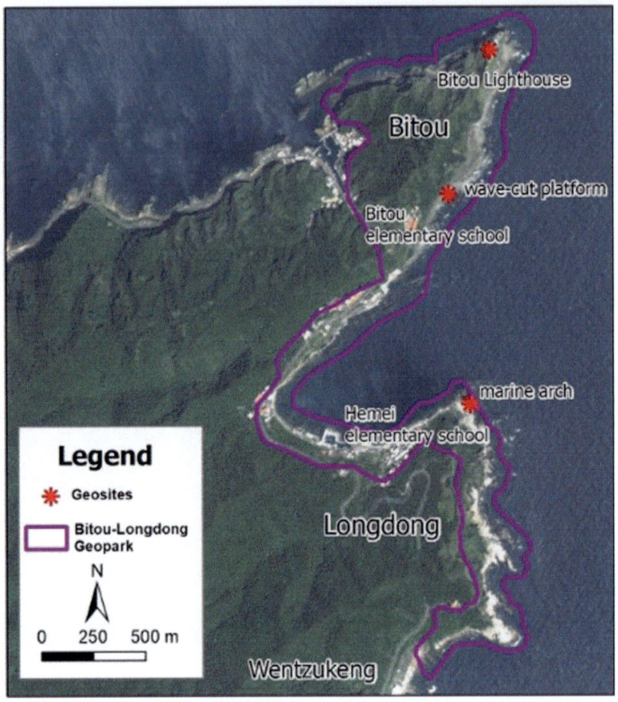

Fig. 1.81 Location map of Northeast Coastal Longdong and Bitou Cape Geopark

Fig. 1.82 Sea cliff landscape (strongly influenced by joints) at Northeast Coastal Geopark

Fulong highlights the finest beach of the geopark as a result of sedimentation of sandstone. Fulong sand sculpture festival is an annual attraction in the summer time, attracts both domestic and international visitors (Figs. 1.89, 1.90 and 1.91). Whale watching around Turtle Island is an experience that combine geological, ecological and human cultures in the geopark.

Guishan Island

Guishan Island literally means Turtle Mountain island in Chinese language. It is so called due to its similarity to the shape of a floating turtle. It is located 10 km east of the Yilan County's coastline. As a volcanic island, Turtle island is relatively young and composed mainly of andesite and

Fig. 1.83 Bird's-eye view of the Northeast Coastal Geopark

Fig. 1.84 Marine cut platform at Northeast Coastal Longdong and Bitou Cape Geopark

volcanic clastic rocks. There are steep cliffs, active hot springs, hills, peaks, erosion caves, lakes, cold springs, and a rich ocean ecology on and around the islet. The island is open to the public with limits for maintaining its carrying capacity. The sulfur emissions and hot springs turned the sea water colors into a visible environmental education for school children. Whale watching is another attraction that people could enjoy around the island offshore with locally operated ferryboats. Turtle Island is a beautiful gem in the ocean for various reasons (Figs. 1.92, 1.93 and 1.94).

The scientific and environmental significance of these landforms is promoted with good practice of environmental

Fig. 1.85 Mushroom rocks on the marine platforms of Bitou-Longdon Geopark

Fig. 1.86 Sea notches in the Bitou-Longdon Geopark

Fig. 1.87 Horizontal bedding at Bitou-Longdon Geopark

Fig. 1.88 Cross-bedding landscape at Bitou-Longdon Geopark

Fig. 1.89 Fulong beach is an estuary of Shuanhsi River

Fig. 1.90 A sand sculpture at the Fulong Sand Sculpture Festival

Fig. 1.91 Panorama landscape of Sand Sculpture Festival at Fulong beach

Fig. 1.92 A view of Turtle island from above

Fig. 1.93 Turtle island releases sulfur which changes the color of sea water

Fig. 1.94 Dolphin near Turtle island

Fig. 1.95 Snorkeling training course at Bitou-Longdon geosite

education and interpretation. The Northeast Costal Geopark is one of the best environmental education sites for eroded coasts. It is the most visited tourist spot along Taiwan's northern coast as well. It is particularly attractive for both children and adults with its aquatic activities and makes the coast gorgeous in the warm seasons (Fig. 1.95).

References

Chen, W. S., et al. (2012). *Together let's make geoparks work*. Taipei, Taiwan: Forestry Bureau of Taiwan. (in Chinese language).

Chen, W. S. (2014). All out of Orogeny. *Geology, 33*(1), 23–25. (in Chinese language).

Chen, Y. H. (2017) A geopark filled with life and death stories. In *Taiwan on geological scale: Tracing Taiwan geology from geoparks* (pp. 148–171). Taipei, Taiwan: Metropolis Publisher. (in Chinese language).

Chi, S. J. (2002). *The conservation guidebook of Tainan's geomorphological landscapes*. Kaohsiung, Taiwan: National Kaohsiung Normal University. (in Chinese language).

Lee, C. T. (2011). *The development process of Tsaoling Geopark*. Taipei: Forestry Bureau of Taiwan and National Taiwan University co-published. (in Chinese language).

Lee, C. T. (2013). *The transformation of Tsaoling's geomorphology*. Taipei: Forestry Bureau of Taiwan and National Taiwan University co-published. (in Chinese language).

Lee, C. T. (2014). Landslides and Tsaoling Geopark. *Geology, 33*(1), 31–35. (in Chinese language).

Lin, J. C. (2011). *Classroom of Taiwan geomorphology*. Taipei, Taiwan: Owl Publisher. (in Chinese language).

Lin, J. C. (2013). *Introduction to Matsu geopark*. Matsu, Taiwan: National scenic area administration of Matsu. (in Chinese language).

Lin, J. C. (2014). Promoting Taiwan Geopark network. *Geology, 33*(1), 20–22. (in Chinese language).

Liu, S. L. (1999). *The cultural and tourist resources of Penghu Marine Geopark*. Penghu, Taiwan: Fishing and Agriculture Bureau of Penghu Government. (in Chinese language).

Su, S. J. (2014). The landscape resources and community participation of Penghu Marine Geopark. *Geology, 33*(1), 26–30. (in Chinese language).

Su, S. J., & Lin, J. C. (2014). The environmental and social implication of geoparks. *Geology, 33*(1), 56–69. (in Chinese language).

A New Way of Understanding Geoparks *for* Society

2

2.1 Natural Setting and Its Significance to Society

In a fast globalizing world, where ideas, technologies and materials flow without much constraint, ways of thinking and actions in society are driven to change drastically from the historical background. It is important to understand and appreciate the stability of the local setting. It is significant because the local environments and cultures are something that local people attach to and the local livelihood depends upon.

With the fast changing environments across the world at various spatial scales, the relations between human society and the natural environment are seen in a new light, where the concept of "environmental ethics" and "sustainability" are of concern and much debated. The consideration of environmental ethics is obviously a result of society's problem of facing a different environmental predicament. Considering environmental ethics becomes a possible solution to the problems created by the uncertainty of the fast changing global environment. In the past, our respect or fear of the natural environment was often significant enough to influence our responses and strategies to cope with natural processes and natural hazards. However, with the development of modern technologies and ways of consumption, natural environments are easily changed or even bought. We can now easily control nature and even destroy environments.

Now with a rapidly changing global climate and environment, societies, at various places and of different spatial levels, endeavor to find sustainable solutions to global environmental issues. These solutions come in many forms, some through bottom-up activities, some seeking to learn from ethnic wisdom and mending ethic relations, and others with the restoration of cultural practices. These efforts aim at examining and reflecting upon local actions to combat the global changes. When facing detrimental natural processes and natural hazards, do the local people know what their strength and weakness are and how resilient their environment and society is? The lucky outcome would be a society of people having the determination to understand their own environment and to learn how to take care of it for a more flexible and adaptable future.

The idea of a global geopark network is a UNESCO initiated and supported global endeavor for a sustainable future; individual geoparks could play a significant role in reviving and conserving our living environmental setting and in supporting civic participations in the environment through learning and understanding. Human society must firmly secure sustainable development and a single definition of that process. It is argued by UNESCO and in this book that we must encourage serious and fundamental local actions to sustain our geological and geographic environments.

There are many topical issues related to geoparks. Issues such as natural resources, geological hazards, climate change, geo-conservation, education, culture, local knowledge, role of women, or science are all significant societal issues and can be pursued by the promotion of geoparks.

2.2 The Core Values of Geoparks and Their Application in Taiwan

The Geopark idea was proposed by the Earth Science Division of UNESCO as a programme in 1999 (document 156 EX/11 Rev.). However, at the 161st session of UNESCO's Executive Board in 2001, the majority of delegates decided 'not to pursue the development of a UNESCO geoparks programme, but instead to support ad hoc efforts within individual Member States as appropriate' (quoted from UNESCO Geopark website, 2015). As a consequence, an ad hoc UNESCO geopark initiative was soon developed by various national states, mainly because of an ambition to promote conservation and sustainable development in economically disadvantaged areas. A Global Geoparks Network (GGN) was soon formed to promote geoparks and their development globally. The long waited formal UNESCO Global Geopark label was ratified in 2015 by the member states of UNESCO. Appreciating the significance of

J.-C. Lin and S.-J. Su, *Geoparks of Taiwan*, Geoheritage, Geoparks and Geotourism,
https://doi.org/10.1007/978-3-030-04895-2_2

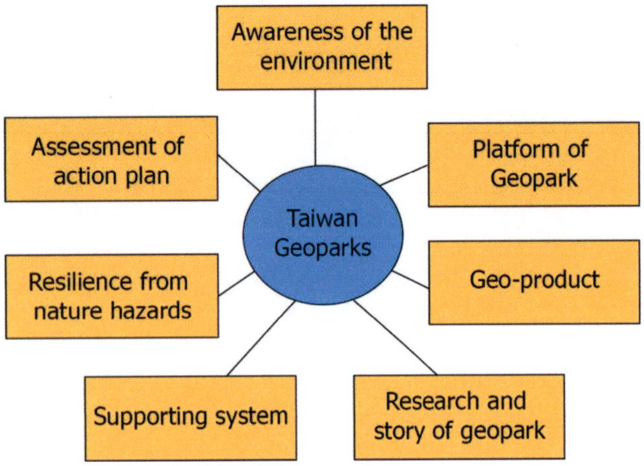

Fig. 2.1 Elements of Taiwan Geoparks

geoparks, Taiwan has come a long way to develop its geoparks with many implied elements, values and methods of UNESCO programme (Fig. 2.1).

The most concerned topics disclosed in the official GGN (Global Geopark Network) website of UNESCO list (as googled in January 2016) in Table 2.1 show a general trend of this rapidly changing environment and society. With some thoughtful reflection, we can translate them into Taiwan's societal concerns or issues that could have implications lead to authentic solutions in each individual geopark of Taiwan or Taiwan's context.

Taiwan, not being a member state of the UN, came to realize the merits of geoparks through the endeavor of its academic society, particularly geographers and geologists. The idea of geoparks was accepted without much resistance; partly because inventories of the geological and geomorphological landscapes of the country have been prepared by Taiwan's central government for almost three decades and partly because geo-related hazards were of much concern in the society. The rare and significant geo-sites, as promoted by Taiwan's geopark programme, are not only significant for scientific reasons, but were regarded as important for environmental education. The Geologic Act of Taiwan was passed in 2010 after a long and painstaking 14-year period of conflicts of interest in the land use of geologically hazard prone areas (Chuang 2010). Taiwan's geology proves to be diverse and hazard-prone in various areas, and thus the standard of building codes and building permission issues at local governments become sensitive and problem-ridden. Environmental values, the need for conservation and respect for nature are vitally important issues, which often conflict with the economic consideration of business people or industry. Land owners, who are afraid of being deprived of the right to their property entitlement, also become voices against geologic legislation that might interfere with their rights. This meant that protection of geologically sensitive areas (that otherwise might be inappropriate to be

Table 2.1 The implications of geopark topics of the Global Geoparks Network for Taiwan's geoparks

Themes concerned for the UNESCO Global Geoparks Network	Implications for Taiwan Geopark Networks
Natural resources	What are newly defined resources: possibilities and limits of resources
Geological hazards	Dynamic environments and hazards: earthquake, typhoon, tsunami, landslides, etc.
Climate change	Dynamic challenges: coastal erosion, water resource, land subsidence, agricultural production, societal adaptation, etc.
Education	Environmental education to prepare people for a changing environment and for better adaptation and adjustment at various levels
Science	Scientific research results with local implications
Culture	Ways of life: cultural heritage and traditional wisdom for maintaining a sustainable environment
Women	Women as human resources, as social inspiration, and as initiators for societal wellbeing; female members involving in local affairs that protect environment, local cultural and social integrity
Sustainable development	Ecology, bio-diversity, coastal land use planning, slope land use feasibility, etc.
Local and indigenous knowledge	Traditional local knowledge of protecting local environment; indigenous knowledge for protecting mountainous and riverine environments
Geo-conservation	Deep understanding of geology for better conservation; landscape inventorying and landscape conservation

developed) could be difficult for investors and developers. Because geological geoparks may create problems of these sorts, the Geological Act provides a means for the various interests to be surveyed and investigated. These surveys aim to provide fundamental knowledge for the promotion of geoparks and sustainable and resilient development, and for the education of the local populace.

Hazard-prone geology is evident throughout Taiwan due to its tectonic situation. However, it was the 1999 Chi-Chi Earthquake (of Richter scale 7.3) and the April 2010 landslide incident at a cut-off slope foot of a dip slope on a northern highway that truly triggered the public awareness of the significance of geology and its relation to hazards and to human society. The promotion of geoparks came at the right time for educational institutions and the public, with national policy support and the local community's keen awareness of their own geo-safety. Soon an Environmental Education Act of Taiwan was passed in 2010 in response to the needs for understanding the environment we live in and we live by. The Environmental Education Act also gives the geopark idea policy support and legitimacy.

The Taiwan political environment and a strong societal demand for environmental soundness and a future of sustainability coincided at this time. The values of geoparks were therefore easily understood and welcomed by the scientific and educational communities. Certain key geopark values become the engine for the development of Taiwan's geoparks particularly at the community level. The core values of geoparks are landscape conservation, environmental education, geo-tourism and community participation. They each have unique connotations, as shown in Fig. 2.2, and are discussed below in the Taiwan context.

Landscape conservation is the first value of geoparks that is easily associated with the scholarly fields of geology and geomorphology. For Taiwan's various levels of governments and academia, the concern of landscape and its value has been in the spot light for the past three decades. The national landscape inventory is constantly being updated. However, landscape conservation needed explanation of its relation and significance to the public. This was partly achieved by teaching and publication of explanatory books and the addition of the term of "landscape conservation" as one of the key concepts for the geography curriculum standards of senior high school education in 2006 amendment. Today, the core value of landscape conservation is increasingly being received by the general public in a favorable way.

A time line for landscape conservation can actually be established combining the landscape inventorying history and a national curriculum transformation. A brief history of Taiwan's landscape conservation will show how it has become standardized in its scholarly fields. Four stages could describe the development of Taiwan's landscape conservation, including the initial, foundation, turn around and restarting stages.

The world now faces global climate change and rapid and mostly unpredictable environmental change, the ambition of geoparks to combine conservation and sustainable development for economically disadvantageous areas has been welcomed. The UNESCO geopark initiative motivated not only academics but also the local communities with special geo-sites who have come to understand their own geological environment and its geographical implications. This understanding has developed respect for the environment and a wish to look after the landscape as an economic and socio-cultural asset.

In Taiwan landscape conservation through Geopark promotion therefore has two particular meanings, including: deeming landscape as a resource and using landscape as a tool for sustainable management. These two tiers of

Fig. 2.2 Values of geoparks and their contextual development in Taiwan

landscape conservation are manifest in Taiwan's promotion of geoparks. Plenty of education programs with geological educational materials at the community and school levels have strengthened the public's view on how landscapes could be of use and how they should be managed for sustainable future.

The national government of Taiwan had undertaken a series of landscape inventory programs for three decades since the late 1980s (Fig. 2.3). The geopark concept has been accepted with ease. Now the geoparks work to ensure that human activity takes its environment into consideration at the level of local communities. Today, Taiwan embraces the geopark concept as being relevant to society in this globalizing and fast changing environment. In 2015 UNESCO formally passed the UNESCO Global Geoparks Program. It is a milestone for human society.

Environmental education through geoparks is another significant value to Taiwan society. With its dynamic environment setting, high population density, and highly developed industrial economy, Taiwan's environment stress is in fact among the highest in the world (Chang 1992; Williams and Chang 2008; Chen and Hsieh 2011; Chen 2013). For that reason, well-planned environmental education for the national school curriculum is vital. Environmental education, however, was not formally a part of national endeavor until October of 1987 when the Taiwan government issued a guideline for an environmental policy framework (Yang 1992). Since then the national Ministry of Education has been concerned about how environmental education could be part of the national school curriculum. Nonetheless due to Taiwan's examination system for entering senior high schools and universities, curriculum and courses not traditionally considered part of the entrance examination were not taken seriously or easily incorporated into formal curriculum by teachers and students. This is why the 2006 amendment of senior high school geography curriculum standards incorporated the term of "landscape conservation" became so important (Fig. 2.4).

It took some time until the national educational transformation in the mid-1990s, when an overhaul of the traditional school curriculum was scrutinized and re-considered. In the new national school education system, mandatory

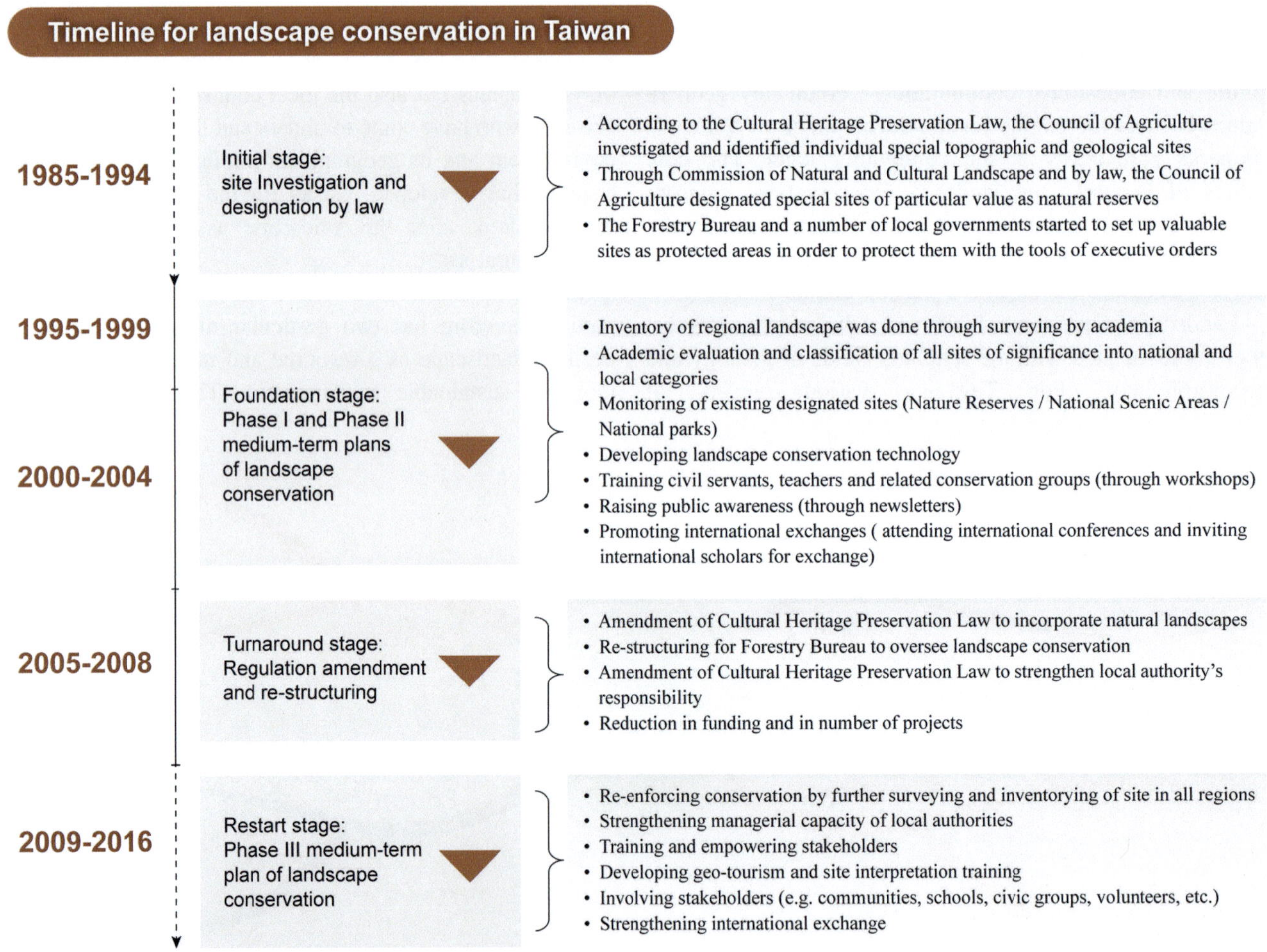

Fig. 2.3 Timeline for landscape conservation in Taiwan

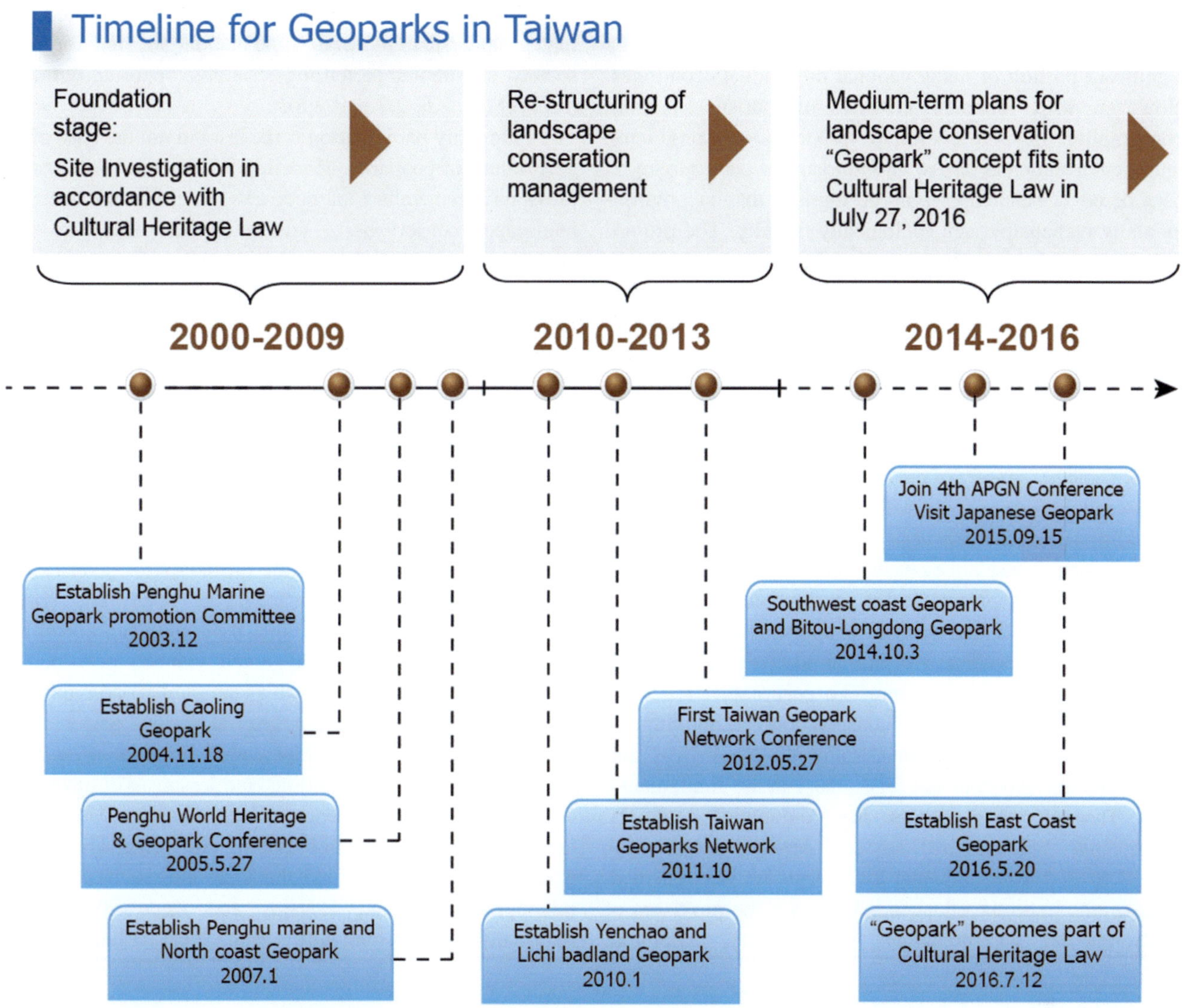

Fig. 2.4 Timeline for landscape conservation in Taiwan

school education is extended to 12 years from an old 9-year system. Environmental education is mandated to be infused into the education of all areas, including language study, arts and humanity, natural science and life technology, health and physical education, and social studies. Since then many aspects of the environment are being incorporated in the curriculum. Further, during the curriculum revision of Geography various forms of environmental education are added, such as in the teaching of industrial development, urban spatial sprawl and regional development. A new core term of "landscape conservation" was even introduced and added in the new geography curriculum of senior high schools. It is a breakthrough of geographical education to extend from knowledge-based to action-based education. With the Environmental Education act passed in 2010 and an alarming rate of increased environmental hazards and incidents of climate change, environmental education was prioritized in Taiwan's education, which makes the geopark promotion much easier. Thus the value of environmental education can be easily conveyed and conducted.

For example, with the education and promotion that a geopark is a designated area with geological heritage of scientific, educational, socio-economic and even cultural significance, geoparks are not only understood as being about geology, but also about geography, culture, ecology, economy, and even areal livelihood. Geoparks utilize such heritage to promote awareness of issues facing our society in the dynamic and fast changing planet we all live on. Thus, geological heritage (geo-heritage) and geological hazards (geo-hazards), geological diversity (geo-diversity), are seen as being intertwined and relevant to Taiwan's landscape conservation and environmental education. For all the geo-tourism is now seen not only a value of a geopark but an opportunity for the local community.

Geo-tourism is a geopark value that coincides with today's word economy where tourism has been considered a significant portion of many national or regional economies. However, with vehement criticism of tourism as being consumptive of non-renewable resources and producing much environmental stress, geo-tourism, or eco-tourism, is considered a balancing form of tourism that is environmentally responsible and ecologically friendly. The promotion of geoparks has the aim of sustainable development. Geo-tourism is therefore automatically related to something sustainable or equivalent to eco-tourism (Fig. 2.5). That is to say, even without a clear meaning or definition of what geo-tourism ought to be, how it should be conducted, or even what the environmental consequences would be, geo-tourism is often conducted towards understanding the geology and geography of the park area. Thus geo-tourism is often used interchangeably with eco-tourism.

Eco-tourism often equates conservation, community and interpretation. These are elements that are also considered in Taiwan's geopark promotion for geo-tourism. Geo-tourism provides economic and societal incentives for conserving and enhancing geo-diversity and protects the natural and cultural heritages of the nature. To ensure that the local community is involved in the conservation act, building the community's capacity to understand the geological and geographical environment is the first fundamental step. Through such process, the community could be empowered and enriched so that personal experiences (Fig. 2.6) and environmental awareness are strengthened. This results in an excellent foundation for the provision of environmental interpretations for visitors. Therefore geo-tourism may be considered as eco-tourism which promotes understanding and appreciation for nature, local society, culture and increasing economic opportunity for the locals (Figs. 2.6, 2.7 and 2.8).

Community participation is the last but not the least of the core values of geoparks. Most importantly it is related to the previous three values of geoparks, in profound ways, as landscape conservation, environmental education and geo-tourism are all linked to the identity of the community. Community participation is significant in many ways. The community and its people is the most fundamental basis for understanding the local needs, perspectives and environmental problems. Community is critical in participatory planning where locally appropriate decisions and plans could be made and conflicts avoided. With positive and beneficial cycles of local participation, just, democratic and transparent ways of making good things happen become possible. Therefore local power can fully prevail and all can contribute to or benefit from the promotion of geopark in a transparent and immediate way. As Scheyvens (1999) pointed out, there are economic, psychological, social and political empowerments to be gained with eco-tourism. This is also true for geo-tourism with community participation.

Eco-tourism has been practiced at various places of the world but not without failure cases. If geo-tourism is to avoid similar mistakes, cautions and lessons will have to be drawn from the experience of eco-tourism. Most cases documenting eco-tourism's failure are cases where local participation is missing and the local interest is lacking or mis-represented (e.g., Cousins and Kepe 2010). Also important is the accountability of the local participants who

Fig. 2.5 Geo-tourism within the geopark is essential for local economy

Fig. 2.6 "Leave no trace" as one conducts geo-tourism

in fact usually need support from some sorts of power. Accountability is not only a sweet burden on the local community but it could be used in a positive and fundamental way as the participants will be given a full chance of gaining in-depth knowledge and information for making sound decisions. In such mechanism, community participation becomes a sure way to achieve with success. In the case of geopark promotion through geo-tourism, the participant community will become a greater human resource than all other values.

In Taiwan, geoparks and geo-sites with unique or scientific values have been incorporated into the Cultural Heritage Act since 27 July 2016. The part of the Cultural Heritage Act designation for geoparks is well constituted with the full participation of the local communities, professionals, academics and government officials of the central and local states. It is also a milestone for Taiwan since the first geopark movements were introduced in 1999. As the core values of geoparks are implemented at various geoparks, different lessons learnt from all geoparks enrich geopark promotion. With all the core values, geoparks can be considered a tool for a sustainable future of Taiwan. Taiwan's sensitive geology and dynamic environment warrants a society keen for conserving its environment and geo-based way of life. The geopark concept provides that opportunity.

2.3 Agenda Setting for Taiwan Geoparks

With the aim of conserving our living environment and being keen to follow international trend, Taiwan's scholarly community proposed a list of 31 potential geo-sites in 2001. Very soon, the Yehliu Geopark located at Taiwan's northern coast and close to metropolitan population was established.

Although it is located within a small part of a national scenic area, the Yehliu Geopark was indeed the first geopark of Taiwan. Even without the full support of the local community, the Yehliu Geopark has been a perfect site for geo-tourism and environmental education to show case interpretation programs for sandstone formations and the various landforms produced by special coastal and biological weathering. Now with its pioneering of an international sharing network among Asian geoparks and advanced environmental education programs, the Yehliu Geopark has become a landmark for Taiwan's Geopark system and its management team a virtuous celebrity. The Yehliu

Fig. 2.7 Mudstone at Yenchao Geopark is very sensitive to heavy rainfall and unsuitable for hikers

Geopark's close proximity to the biggest metropolitan area of Taiwan is a plus for its promotion. It is a significant geopark for international networking and conferencing where a lot of experiences have been exchanged. The Yehliu Geopark has been enjoyed by many visitors each year (Figs. 2.9 and 2.10).

In late 2003, a county committee to promote a marine geopark in Penghu was established. Penghu county, being an off shore archipelago, was the first county of Taiwan to establish a Geopark Commission at the county level. In 2005, with the support of the Forestry Bureau (responsible for national nature conservation), the Taiwan national government and the Penghu county government held an inaugural international conference in Penghu to promote Taiwan's geoparks. Entitled "World Heritage and the Geoparks of Taiwan", the conference was attended by geological and geographical scholars and experts of related fields from around the world following discussion and field trip observation, all participants recognized the scientific and educational significance of the geological environment of the Penghu Archipelago. All agreed that the landscapes and many communities of Penghu were ready for the establishment of a geopark. The Penghu county government appointed a geopark committee in the same year and zealously advocated a "Penghu Marine Geopark". It was the first initiative of the Taiwan government and society to promote geoparks as a means of maintaining a sustainable future for the country.

With the successful experience of the Penghu archipelago conference and field observation and some later studies, various initiatives were made at other parts of Taiwan, such as Yenchao and Lichi. The sites chosen are sites selected from the national inventory of geo-sites that started national surveying in the late 1980s. For instance, in 2004 the Tsaoling Geopark was designated by a coalition of the local community and the scholarly community due to its dynamically changing landscapes and its significance in the education for hazard-prone environment.

As some successful experience and beneficial insights were gained from communicating with the locals on

Fig. 2.8 Yenchao Geopark is good for geo-tourism and environmental education

Fig. 2.9 Yehliu geopark is famous for differential erosive landforms

updating and changing geo-sites, the time was mature for a national even and declaration for geopark. First, a logo for Taiwan's geoparks was identified and selected through a national competitive call for geopark logo which then becomes the logo of Taiwan geoparks in general (Fig. 2.11). The shape of the logo represents the boundary shape of Taiwan's main island. Its colors show the diversity of landforms of Taiwan. The green color represents the western fertile plain, and the brown color the east rocky coasts and its earthy and honest people. The two strokes of the letter G in the center of the logo indeed represent Taiwan's tectonic backbone. The blue color showcases the Snow Mountain Range and the Jade mountain range where the highest mountain of Taiwan is located. The second stroke of the letter G of the dark brown color illustrates the backbone of Taiwan, the central mountain range. The logo has been so

Fig. 2.10 Much appreciated Queen's head rock is a typical mushroom rook at Yehliu geopark

much fond of by the geopark community that some of its variation was adopted by other geoparks. The case of Penghu Marine Geopark is one with the additional boundary shape of Penghu archipelago to the west of the Taiwan Geopark Logo to showcase its site and location (Fig. 2.12). As it is a case of showing pride and identity, people will certain not mind its disproportionately large size of the archipelago.

In October of 2011, under the umbrella of "*Act now and unite all to benefit future generations*", an international conference for landscape conservation was held in Taipei, Taiwan. A total of 150 individuals from around 12 countries came to participate in Taiwan's national declaration for geoparks. It was when the "Landscape Conservation Taipei Declaration 2011" was affirmed and became a foundation for geopark promotion in Taiwan. The 2011 Taipei Declaration was signed into effect for the promotion of geoparks (Fig. 2.13). For promoting effective and supportive communication and learning among all geoparks and their facilitators, the Taiwan Geopark Network was established at the same time, endorsed by the Forestry Bureau and the Tourism Bureau of Taiwan. Six geoparks were identified and formally established in the conference, including the Penghu Marine Geopark, the North Coastal Yehliu Geopark, the Tsaoling Geopark, the Yenchao Geopark, the Lichi Badland Geopark, and the Matsu Geopark.

With their own unique environments and characteristics, the six areas were assessed and made into geoparks. After this time, the number of Taiwan's geoparks has increased to

Fig. 2.11 The Logo of Taiwan Geoparks and Geopark Network

Fig. 2.12 The logo of Penghu Marine Geopark that adds the archipelago boundary shape to the Taiwan Geopark Logo

nine, the new ones including the Southwest Coastal Wetland Geopark (Yun-Chin-Nan Geopark), the Northeast Costal Bitou and Longdong Geopark (Bitou-Longdon Geopark), and the East Coastal Geopark. Each has distinguished valuable in geology, education, ecology and environment.

Further, the landscape within each geopark has close connections between local life and livelihood and geological features. The Penghu Marine Geopark features basaltic landscape and marine ecology. Tsaoling Geopark features dynamic landslides and structural landscapes. The Yenchao Geopark is characterized by mud volcanoes and badlands. The Lichi Badland Geopark features mélange mudstone badlands. The North Coastal Yehliu Geopark contains erosional sandstones. The Matsu Geopark is famous for its granite, cold war heritage and cultural landscapes. The Southwest Coastal Wetland Geopark is characteristic of wetland and swampy landforms with bio-diversity. The Northeast Coastal Bitou-Longdon Geopark has characteristic sandstones and marine platforms dominated by fault line and joints. The East Costal Geopark is a park echoing many other coastal landforms but with plenty of evidence of uplifted tectonics. All of these landscapes are rare, unique and important for environmental education and scientific reasons, and are fundamental assets for local culture and socio-eco-systems.

2.4 Concluding Remark

Geoparks have become a frontier where the physical environment and the human aspects of life are highly integrated, and the geopark practices are linked to realistic ways of life. Geoparks also signify that our human society has come to the realization that ways of life are not only shaped by the planet, but also the forces for conserving the planet. Therefore, understanding how geoparks advocate "geo-science *for* society" will lead us to the thought of the utility of knowledge for human society. In the promotion of geoparks in Taiwan, it is beneficial for the scholarly community to realize how much geological and geographical knowledge could be of use for the local communities to reflect upon life and livelihood. Vice versa, the scholarly community is able to reflect upon how knowledge can be directed into products that will provide services of use to the local communities whose ways of life ad livelihoods are crucial. With love of their own land and life, further understanding and appreciation of maintained sustainable environments for future generations becomes prioritized.

Under the Cultural Heritage Act, the Forestry Bureau has investigated and identified individual special topographic and geological sites since 1980s. Through the Commission of Natural and Cultural Landscape and by legislation, further endeavors have been made for sites of special values, such as the designation of natural reserves and geo-sites. Further, with the 2016 amended Cultural Heritage Act in which geoparks are particularly singled out for conservation purpose, the Forestry Bureau and a number of local governments which previously supported or set up geoparks, are strengthened for its support of geoparks for sustainable development. With the public understanding of the significance of conservation and the willingness of the locals to conserve their own environment, the geopark communities have been very promising.

The national inventory of regional landscapes was produced by academia with support from national government. Academic evaluation was followed by classification of all sites of scientific and educational significance into national and local categories. The program also monitored existing designated sites and developed landscape conservation technology. By training civil servants, teachers and all related and enthusiastic conservation individuals and groups to raise public awareness, the concept and techniques of landscape conservation have taken root in Taiwan quite positively and is a good development of geo-science *for* society.

Fig. 2.13 The Banner of Taipei Declaration of Landscape Conservation, 2011 signed by participants and community representatives

Due to the dynamic of Taiwan's geological environment, constant surveying and inventorying of valuable geo-sites and geoparks in all regions are crucial. As knowledge improves, the managerial capacity of the local authorities is improved. The training and empowerment of all stakeholders has been a major framework of geopark promotion in Taiwan. The values of geoparks are instrumental for Taiwan's geoparks, and they contribute to the successful endeavor in Taiwan.

References

Chang, C.-Y. D. (1992). A study of environmental stress and political conflict over slope land development in Taiwan. *Global Bioethics, 5* (1), 43–50.

Chen, H.-S. (2013). Empirical applications of an environmental stress indicator and the environmental efficiency revolution in Taiwan. *Environmental Skeptics and Critics, 2*(3), 82–89.

Chen, H.-S., & Hsieh, T. (2011). Empirical research on constructing Taiwan's eco-environmental stress indicators system. *African Journal of Environmental Science and Technology, 5*(4), 337–347.

Chuang, Y. S. (2010). *After a long 14 years—Finally comes the geology act.* http://www.digitimes.com.tw/tech/dt/n/shwnws.asp?cnlid=90&cat=320&cat1=20&id=0000180577_X706PIV56ZOLOT8CGQZ4G (accessed 2016; written 4-29-2010 DIGITIMES) (in Chinese).

Cousins, B., & Kepe, T. (2010). Decentralisation when land and resource rights are deeply contested: A case study of the Mkambati eco-tourism project on the wild coast of South Africa. *The European Journal of Development Research, 16*(1), 41–54.

Scheyvens, R. (1999). Ecotourism and the empowerment of local communities. *Tourism Management, 20,* 245–249.

Williams, J., & Chang, C. D. (2008). *Taiwan's environmental struggle: Toward a green silicon island.* New York: Routledge.

Yang, K. C. (1992). A brief development history of environmental education. *Quarterly of Museology, 6*(3), 3–10. (in Chinese).

An Island State on a Cradle

3

The cradle upon which Taiwan's socio-cultural civilization has been shaped is influenced by the way its people perceive the natural environment. Situated on the convergent boundaries of the Eurasian continental plate and the Philippine Sea Plate and located on the Pacific fire ring, Taiwan has regularly experienced constant sensible and non-sensible earthquakes (Fig. 3.1) (TCGS 2014a, b; Lin 2012).

Active faults are particularly critical in analyzing the distribution of areas easily affected by earthquakes. These geologically natural phenomena profile not only Taiwan's economic activities, but also its socio-cultural civilization in many ways. Economic, social, political, architectural, and even educational activities are just some aspects of life heavily shaped by earthquakes (Fig. 3.2).

The area of the major island of Taiwan is only 36,000 km^2. With a north-south length of no more than 400 km, and east-west length of less than 150 km, Taiwan's main island has a geological structure resembling the shape of the island. Figure 3.3 reveals Taiwan's geological conditions in which the major layers go northeast-southwest or north-south. With the highest peak of 3952 m (Mt. Jade, or Yu-Shan in Chinese language; the holy mountain of Taiwan) located in the central mountain range of the island, the slopes of Taiwan are very steep. Less than one third of its area is flat and rivers mostly have steep gradients, making some catchments hazardous in the event of heavy rainfalls. The relief image of Taiwan island shows the distribution of altitude and reveals that plains are limited to the west of the island (Figs. 3.3 and 3.4). Furthermore, the geological and geomorphological foundation has constrained significantly the areas where major populations reside and make their living. How the life of the island population could be influenced by such natural setting is a basic knowledge that is worthy understanding in order to develop the philosophy of getting along well with the nature.

Weather and climate are another set of elements interacting with the natural and human characteristics and life of Taiwan. In particular, the annual and seasonal typhoons from the South Pacific Ocean and the Philippine Sea are tropical storms that bring tremendous rainfall and high wind and become detrimental to island life. The weather conditions have shaped Taiwan's life in many ways. It could easily be revealed how the island state has been influenced by typhoons and a map of major typhoon routes over five decades is shown (Fig. 3.5). On average each year 3–4 typhoons hit Taiwan, and there are nine major routes of typhoons landing on the island. The places where typhoons land are mostly damaged, but it does not imply that places far away from the landing points of typhoons are safe due to anti-cyclone wind and rainfall. With the small territory of Taiwan, sometimes a typhoon would cover the whole island of Taiwan and damage it tremendously.

With the particular geological and weather conditions, Taiwan's natural state is determined. This is why the Taiwan society has to be adaptable and accommodating to its natural setting to make an environmentally resilient society. The response and adaptation of the people to these conditions act as a significant control of Taiwanese society. This is the true meaning of an environmental 'cradle' (Lin et al. 2013b).

3.1 Tectonics and Geology: Their Implication for Taiwan

The conditions of the natural environment allow Taiwan's contemporary society to become what it is today. This chapter will focus on Taiwan's geology, geomorphology, climate, ecology, socio-cultural milieu, economic development, and community development in order to allow readers a better understanding of the motivation for the geopark development.

Taiwan is an island state and located at the colliding junction of the Eurasian Plate and the Philippine Sea Plate (Fig. 3.1). The northwest-wards movement of the Philippine Sea plate (Fig. 3.2) forms a constantly tectonic impact to the

J.-C. Lin and S.-J. Su, *Geoparks of Taiwan*, Geoheritage, Geoparks and Geotourism,
https://doi.org/10.1007/978-3-030-04895-2_3

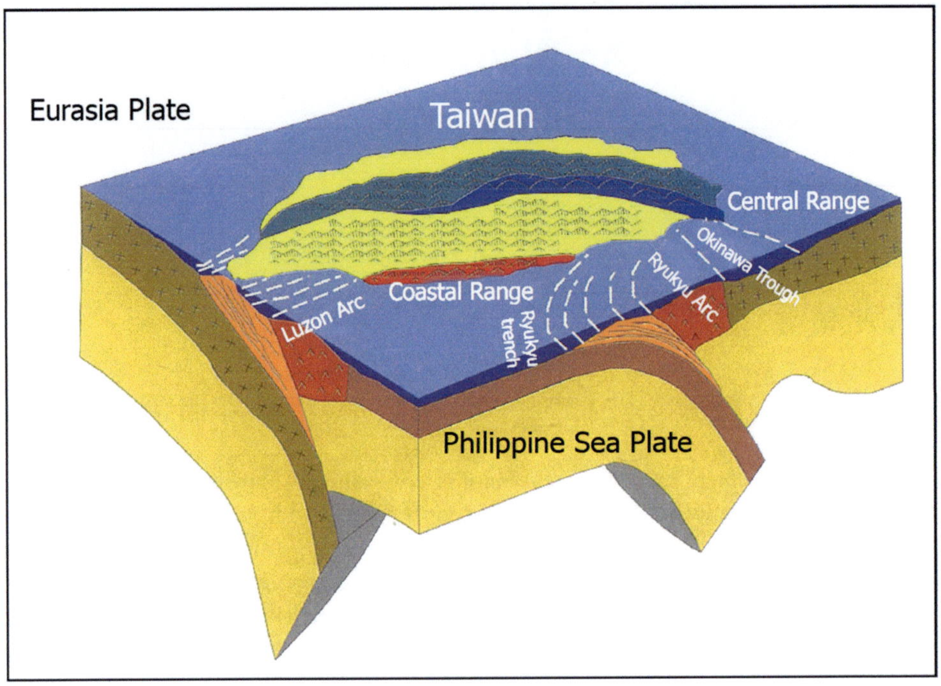

Fig. 3.1 Tectonic plate movements around Taiwan

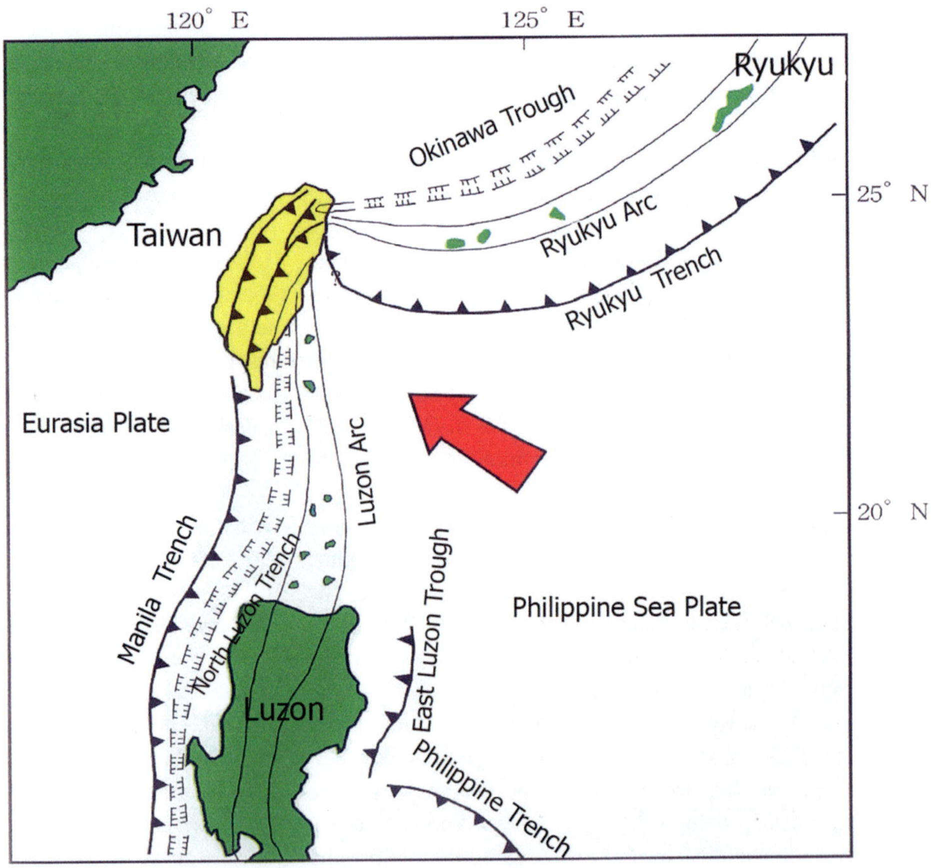

Fig. 3.2 The tectonic setting of Taiwan

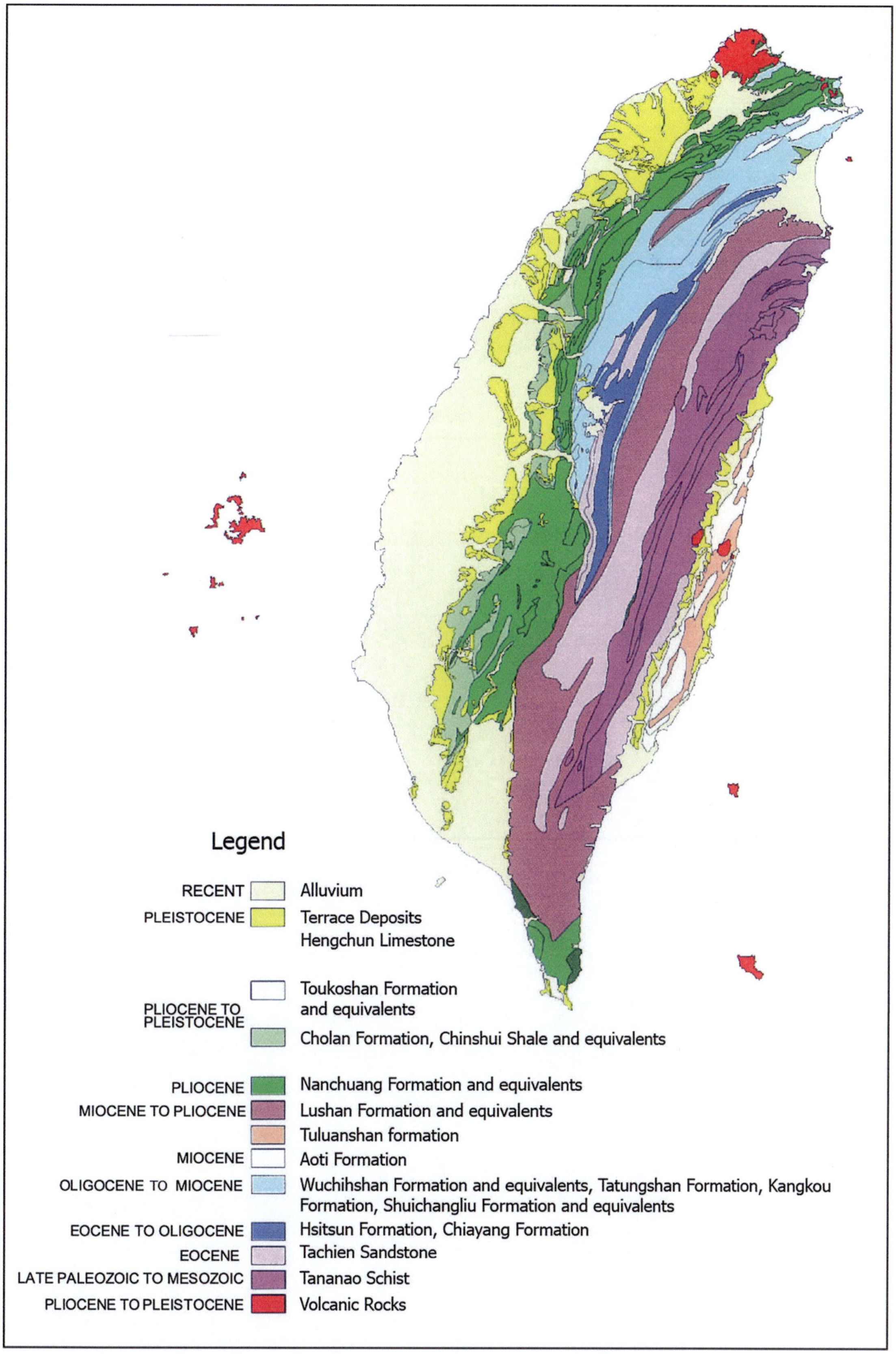

Fig. 3.3 Geological map of Taiwan (*Source* redrawn from Taiwan Central Geological Survey)

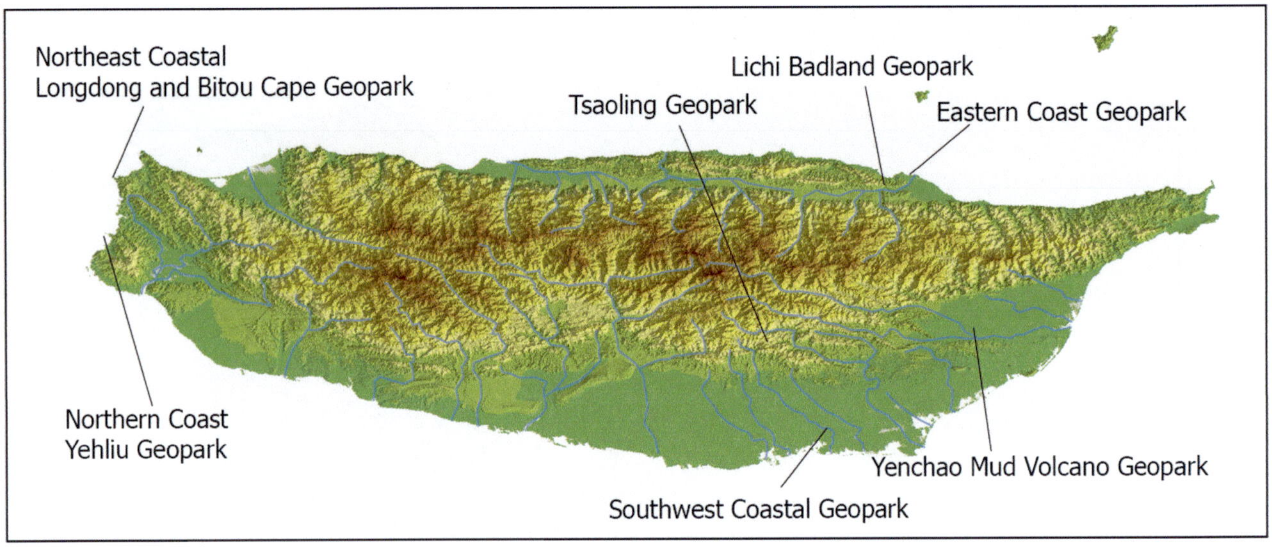

Fig. 3.4 Relief image of Taiwan Island

Fig. 3.5 The major directions of typhoons attacking Taiwan

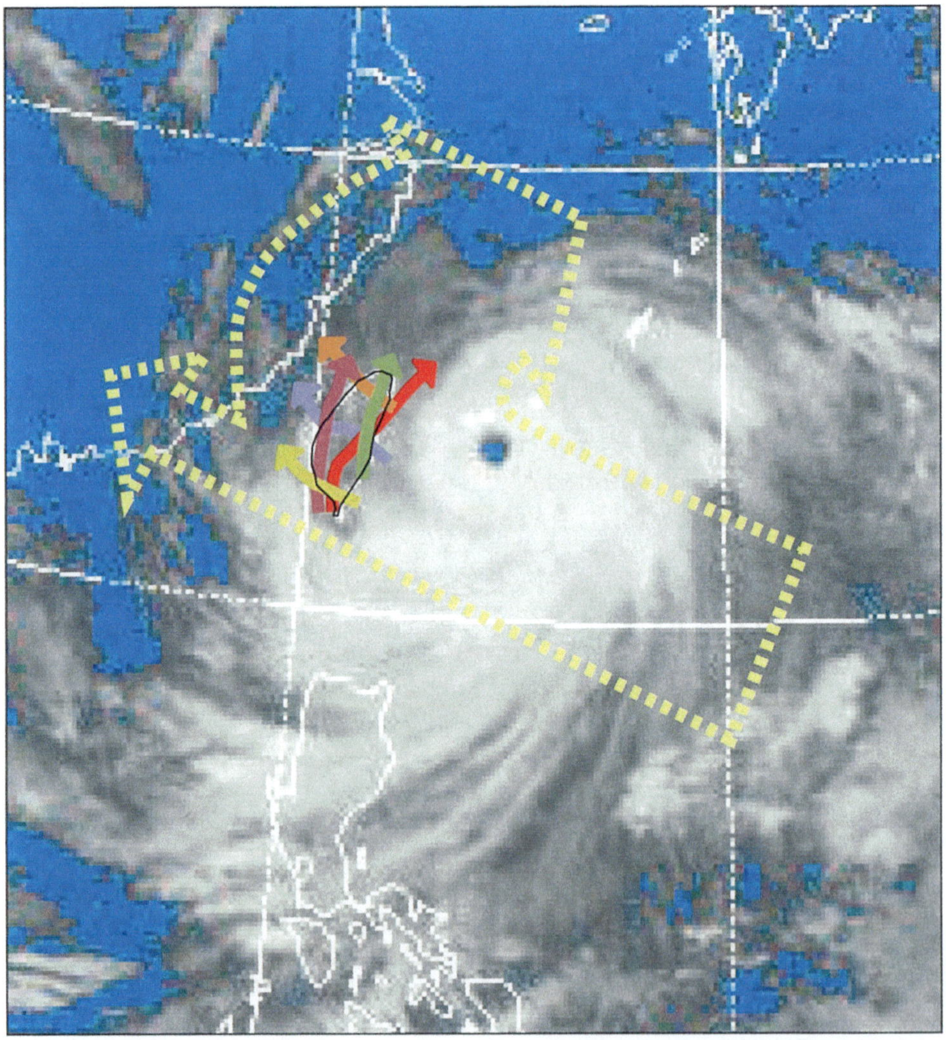

Eurasian Plate to its west, and is associated with frequent earthquakes along active faults on Taiwan Island (Figs. 3.6 and 3.7). That, in turn, distinguishes Taiwan as a unique, vulnerable island land mass and landscape which is prone to landslides and other hazards related to geological and geomorphological processes. This collision of a sea plate and a continental plate also causes Taiwan's constant uplift at a rate of 2–4 cm per year (Fig. 3.8). The region in which Taiwan is located is one of the most significant orogenic belts in the world. Due to the collision of the continental and ocean plates and its geological characteristics, the landforms, in particular riverine landforms, of Taiwan are fragile and susceptible to processes of riverine erosion such as detachment, transportation and deposition. With frequent seismic activities, strong geomorphological processes prevail thus geo-hazard becomes a term that Taiwan's scholarly community and public are familiar with.

The combination and interactions of subduction, sedimentation, metamorphism, and the uplifting of land plates for an extended period of time has created the now complex

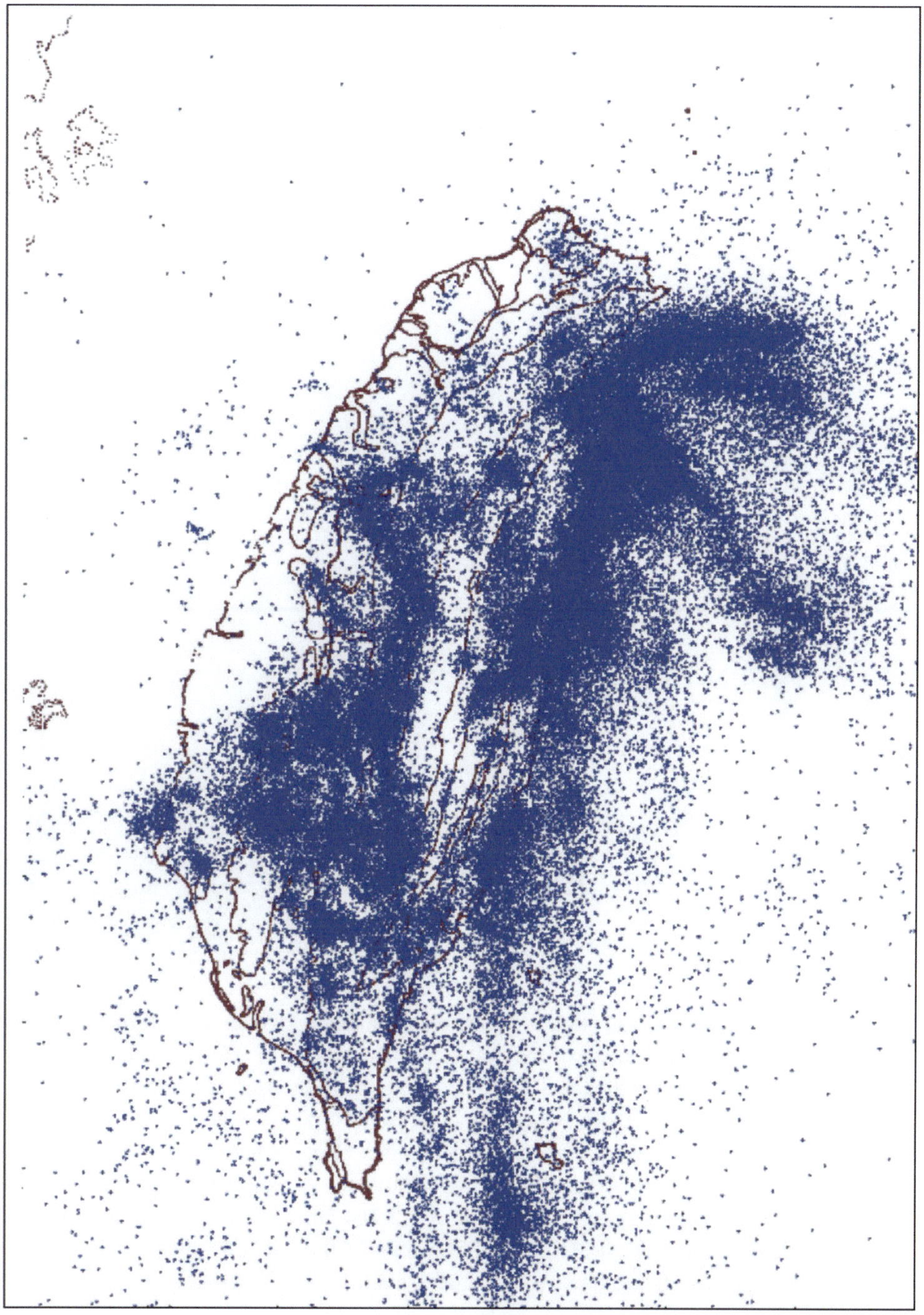

Fig. 3.6 Earthquakes in and around Taiwan during the last 100 years (*Source* Central Weather Bureau, Taiwan)

Fig. 3.7 Distribution of active faults in Taiwan (*Source* Taiwan Central Geological Survey)

geological structure of Taiwan (Fig. 3.7). Different methods to delineate the speed and effect of the collision have been applied, as an average movement 8.3 cm per year can be estimated (Camanni et al. 2014). Numerous other effects of such constant collision are provided by geologists. For example, the rate of uplift of the Heng-Chun Peninsula—determined by 14C dating—could be as high as 0.2–0.4 cm per year. It is often speculated that the uplifting rate of the Central Range could be several times more than this. According to more modest scientific estimations, the island of Taiwan could have been uplifted more than 50 m per 10,000 years (Fig. 3.8).

Due to the constant collision of the plates, mostly unnoticeable at daily life, Taiwan's landforms are fragile and susceptible to processes of erosion and weathering, such as landslide, transportation and deposition. Like a sculptor with sharp tools, these geological processes carve the top soft material of countless hills, mountains and valleys. There are

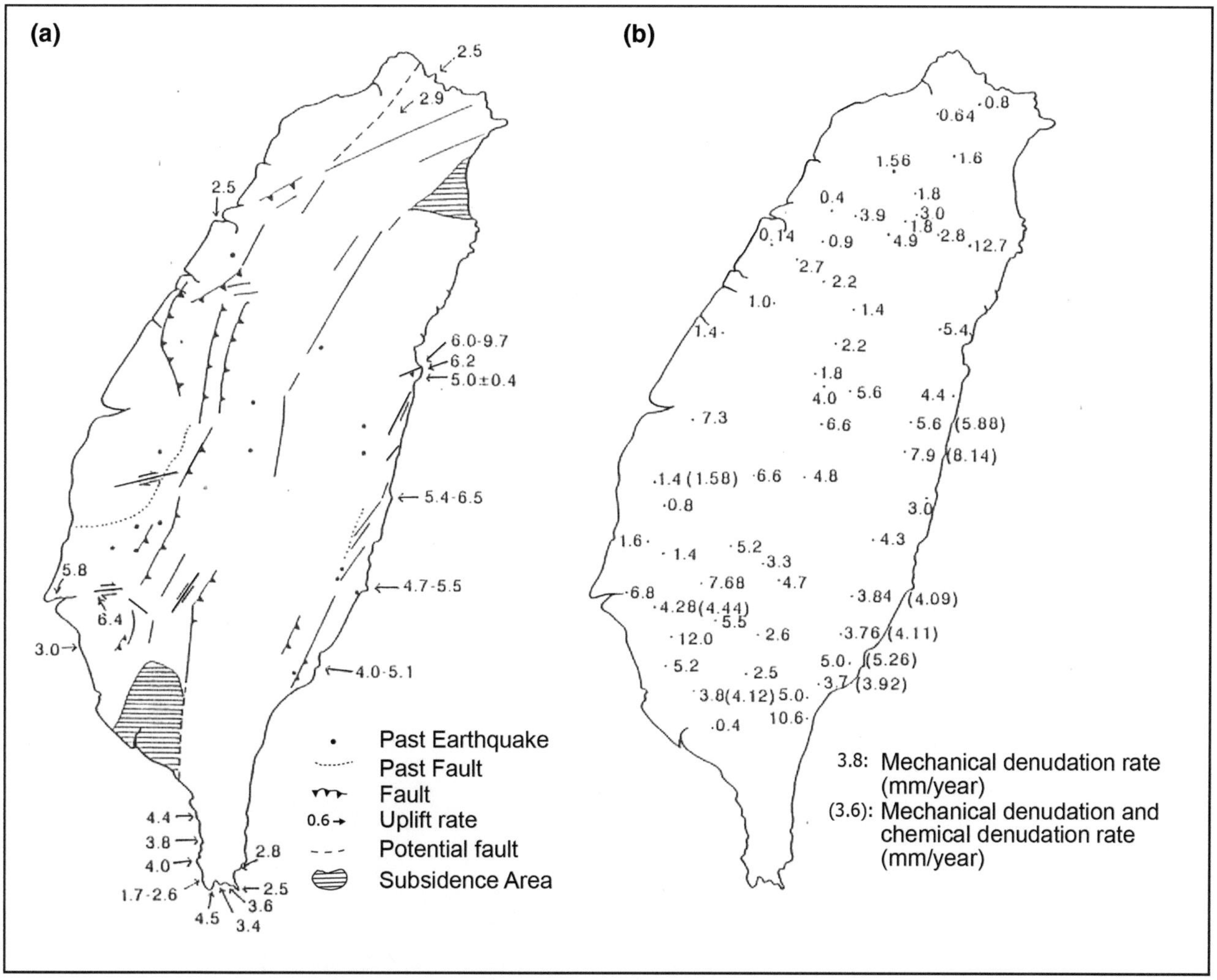

Fig. 3.8 Uplifting rate of Taiwan's land mass

more than 250 mountains in Taiwan that are over 3000 m in height. Rivers cutting through all these hills and mountains have been momentous forces for landscape formation. Thus, 24 major rivers are monitored and managed by the central government, as well as 92 general rivers supervised by various county governments. In respect of these diverse river management systems, the civic society's knowledge and observation of rivers and river changes often play a substantial role as voluntary information provider. In all the high mountains and low valleys, all the rivers occupy two thirds of the small island of Taiwan of only 36 thousand km^2. Essentially every life and life experience is easily related to rivers. However, the geologically dynamic and geomorphological forceful rivers of Taiwan could be challenging and potentially hazardous (Chyi 2006), rivers are not nature to be taken lightly.

Taiwan's last major uplift took place two to four million years ago between the Pliocene and the Pleistocene period as the island mass collide with the Eurasian continent. Due to this (the Penglai Orogeny) and the subsequent movements of the earth's crusts, the island of Taiwan rose from the ocean. Therefore, it is often said that Taiwan's tectonic condition is relatively young.

The geo-diversity of Taiwan is magnificently rich and outstanding landscapes with special value can be treasured everywhere. However, for Taiwan's landscapes to be sustained and to secure the landscape's meaning to community life and livelihood, appropriate conservation and local governance are necessary. Particularly important is the coordination among the national policy structures, various levels of governments and the local communities. Such coordination creates a social system that understands how civilization is built around and articulated between the natural conditions and the human society. To create a good practice of landscape conservation that will take care of the ecosystem for a sustainable human society, it is necessary to understand how nature is integrated into the society's cultures and ways of life. Vice versa, how the ways of life of the human society influence the status of the natural environment is significant for our adjustment strategies for a sustainable future. This rationale underpins the significance of "geopark" promotion, as it emphasizes science for society.

3.2 Climate Induced Hydrological and Geomorphological Processes

Taiwan has a subtropical climate and is located within the East Asian monsoon belt. Annual precipitation is mainly from the frontal rain between the late spring and the early summer, typhoons and seasonal northeast monsoons. The average annual precipitation of Taiwan is more than 2000 mm, with a varied distribution across the island (Fig. 3.9). Mt. Jade, the highest peak of Taiwan, has an annual rainfall of approximately 3000 mm. The windward area in Hou-sao-liao of North-east Taiwan receives an average of 3700 mm of rainfall per year. The Penghu archipelago, off-shore to the west of Taiwan Island in the Strait, has the lowest rainfall of about 1000 mm per year. Precipitation is significant to Taiwan's economy, but is often the source of geo-hazards as well, which will be discussed later.

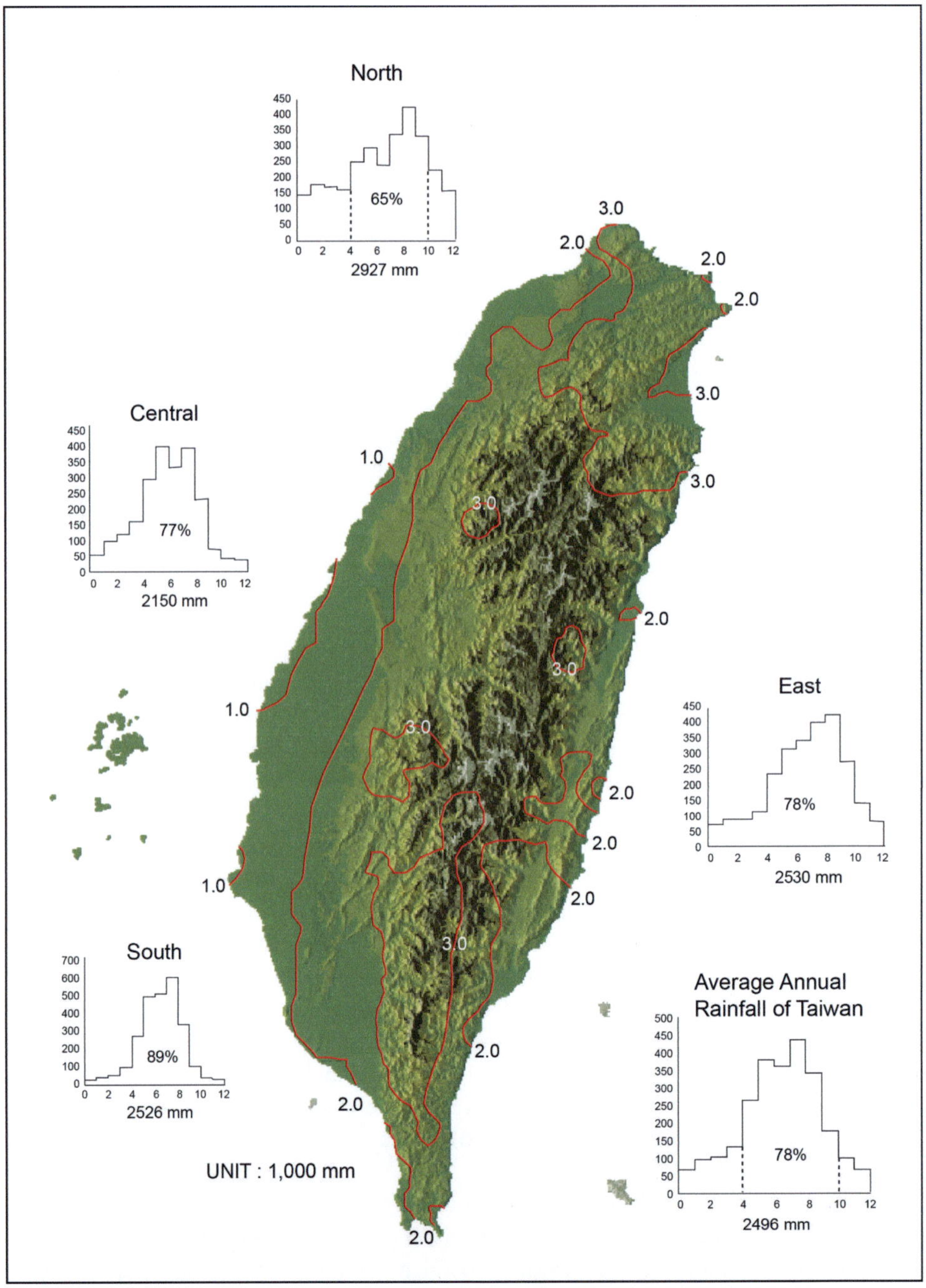

Fig. 3.9 Distribution of Precipitation in Taiwan

Fig. 3.10 A scene of landslides caused by the detrimental Chi-Chi Earthquake

Northern Taiwan is within the windward area of the winter Asian monsoon. Monsoon and winter cold fronts often bring considerable rainfall for the island's northern part. With the winter precipitation, northern Taiwan's monthly rainfalls are higher than those of southern Taiwan. Southern Taiwan has relatively low precipitation, particularly during October to April each year. The precipitation of Taiwan is spatially uneven and seasonally unstable. With global climate change and the increased numbers of extreme climatic events, Taiwan's landforms and consequently its society and settlement that are depending on a stable environment are facing strenuous challenges.

Heavy rainfalls brought by typhoons are often close to world records. For instance, Typhoon Herb in 1996, recorded 1748.5 mm of rainfall within 24 h of time, resulting in injurious landslides in the central mountainous area. More than 500 housing structures were destroyed and 50 people became casualties. Typhoon Herb was the first typhoon that brought awareness of the compounding effect of heavy rainfalls and landslides to the general public in modern hazard history. Typhoon Morakot in 2009, one of the worst typhoons, recorded nearly 3000 mm of rainfall in 4 days. The casualty caused by Typhoon Morakot was beyond description and many lessons were learnt hard and heart breaking. Hsiao-lin village, a remote hill side settlement of the southern city of Kaohsiung was totally erased over night by heavy rain which triggered floods and landslides. A loss of 474 lives in one detrimental and devastating hazardous event saddened the island state. Tremendous damage and environmental destruction were witnessed in the mountain areas and hill sides during Typhoon Morakot. Rains caused by typhoons have taught the islanders of the fragile landforms and the significance of environmental conservation (Figs. 3.10 and 3.11).

The average number of typhoon encountered in Taiwan, since the late 19th century, is 3.8 per year. Typhoons often bring abundant rainfall and causes rapid and dramatic erosion and consequently creates floods and loss of life and property. Typhoon disasters are often at their worst with the compound effect of extreme rainfalls. hydrological features and fragmented geomorphology (Lin 2012; Lin et al. 2013a). As the subtropical climates become more erratic due to unpredictable global climate change, the hydrological and geomorphological processes triggered by climate will become even more difficult to predict. Even worse is that the environmental system will not distinguish natural from human phenomena. For instance, landslides in the mountainous areas block roads and cause the rise of river water level to break or destroy bridge abutments and river embankments, so flooding of settlements occurs. All economic and socio-cultural activities at settlements are consequently severely curtailed.

Typhoons not only bring wind and rain, but also intrusion of salty sea water into the plains where agricultural and aqua-cultural undertakings are mainly located. This shows again that natural phenomena become hazardous to human society. Both cases relate to flooding but with various causes and consequences. Therefore, the way how the human society faces the natural environment and its characteristics and effects needs a very broad perspective to create new horizons to come up with appropriate philosophy and practice of managing the nature.

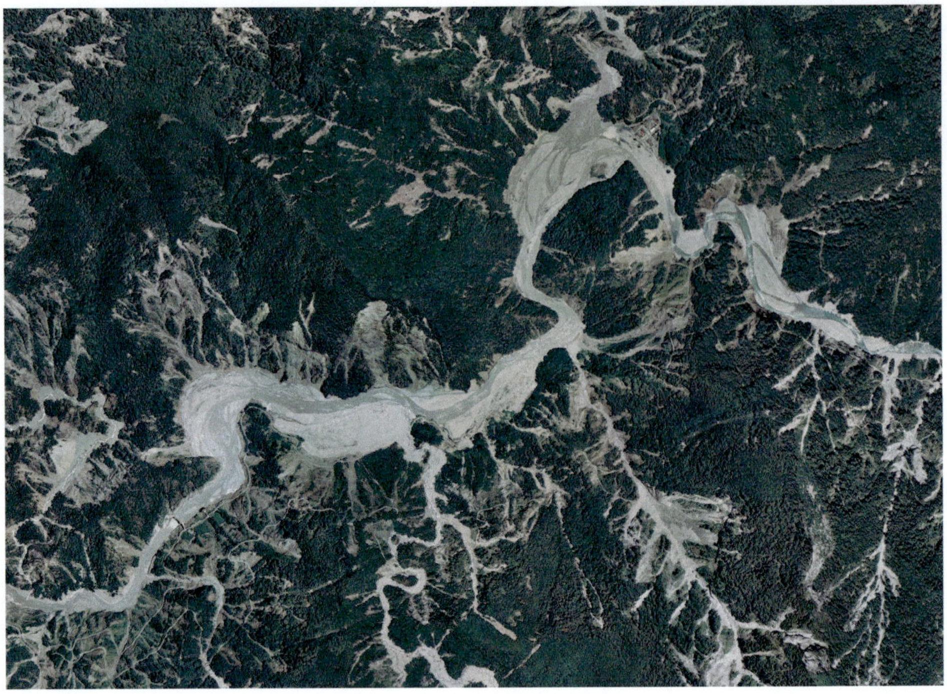

Fig. 3.11 A 30-m high debris pile on Dajia River channel

Geomorphological processes

The combination of crustal extrusions, unique bedrocks and severe weathering processes, results in a large diversity of landforms in Taiwan that reflects on one hand the geomorphological processes, and demonstrates, as well, strong surface erosion with exuberant transportation and deposition. From the almost 4000 m high mountains to the terraces almost at sea level, Taiwan's geomorphology and landscape is a natural classroom for the dissemination of knowledge and understanding of orogenesis. It is a perfect field learning laboratory for developing skills for conservation, as well as disaster mitigation and reduction.

External geomorphologic processes are very dynamic in Taiwan. Typhoons and heavy rainfalls are significant factors in the erosion and weathering process. The subtropical warm and humid weather accelerates physical and chemical weathering. Many slope failures which occur due to poor

Fig. 3.12 Sediment deposit on Kaoping River channel after Typhoon Morakot

land use on geologically unstable areas, are often compounded by extreme weather events. For example, the Chi-Chi earthquake of 1999 in central Taiwan triggered a 30-m high debris flow to the lower channel of Dajia River (Fig. 3.11). The earthquake mobilized new materials from the slopes, the cultivated land and the accumulations from past weathering events. Another case is the huge sediment deposition in the Meishan area of the Kaoping River due to heavy rain as a result of typhoon Morakot in 2009. The typhoon triggered debris flows which piled up to a height of 60 m on the riverbed of Kaoping River (Fig. 3.12).

These cases point to the extremely unstable nature of Taiwan's rivers and river catchment areas. They demonstrate the significant role of the subtropical environment in the geomorphological processes due to weathering and erosion (Lin et al. 2013a). Such geology- and climate-driven geomorphological processes cannot be easily controlled by human and societal intervention. Sometimes human interference is even worse and creates additional loss and hazards. However, with proper local community's conservation knowledge and observation skills, the loss and human induced hazards could be mitigated or even largely reduced.

3.3 Geo-diversity

With the compound effect of a sensitive geological framework, tectonic movements, typhoons, monsoons, and subtropical climate, Taiwan's landscapes are bountiful and entitled to the label of rich "geo-diversity". The diversity of geology and geomorphology of Taiwan is critical to the cultural diversity and bio-diversity of Taiwan. It is obvious that the ways of life, cultures and livelihoods of all geoparks of Taiwan are intrinsically connected with, sometimes conditioned or restricted, by geological and geographical conditions.

With the knowledge and understanding of each geopark's local conditions, geopark communities are able to develop sustainable goals for themselves and their environment. In the following, selected landscapes are illustrated:

3.3.1 Geo-diversity

The particular geological condition of Taiwan generates many typical tectonic landforms. A rising land mass with

Fig. 3.13 Cuesta landscape at Yehliu geopark

active faults, constant noticeable or unnoticeable earthquakes, and all sorts of volcanic features, characterizes the diversity of Taiwan's landforms and its geomorphology. Some features turn out to be hazardous, although some are so breathtaking that they attract visitors. Cuestas are just one example of beauty (Fig. 3.13).

3.3.2 Mudstone and Badland Landscapes

A narrow strip of 1200 km^2 of mudstone landscape is located in south-western Taiwan. Mudstones and shales are made of silt- and clay-sized particles. Their apparent difference is that, when dry, mudstones break into blocky pieces and shales break into thin chips with roughly parallel tops and bottoms. Mudstones are easily eroded by torrential rains and become badlands. Mudstone is fragile and mostly barren with little capacity to support vegetation or intensive agriculture. Each year between October and April when the weather of southern Taiwan is mostly dry, the surface of the mudstone shows cracking and little signs of life (Figs. 3.14 and 3.15).

The mudstone area is home to mud volcanoes or mud domes. These features are created by the ejection of liquids and gases that originate beneath the earth's surface. Methane makes up about 80% of the gases released. Erupted materials are watery slurry which may contain hydrocarbons and is acidic or salty. The drying processes result in unique surface patterns and landscapes that are of particular interest to visitors. There is a total of more than 60 mud volcanoes found in the southern cities of Tainan and Kaohsiung and in Pingtung County (Figs. 3.16 and 3.17) (Chyi 2002).

The mud volcanoes of Wushanding are active and new mud volcanoes are formed occasionally. The mudstone area of Moon World in Kaohsiung often has steep slopes of 45°. In dry seasons, the hardness of the rocks makes it easy to climb and walk on, but during rainy seasons they become impossible to even stand on. Sediments found in the gullies are often materials from the previous rainfall events. The gullies are so well developed that they form regular patterns and landforms that are devoid of vegetation. This landscape demonstrates the inter-relationship between geology, topography, soil, plants and water, and is a natural laboratory for learning about flora, fauna and conservation.

Fig. 3.14 A mud volcano at Wushanding, Yenchao Geopark

Fig. 3.15 Mudstone is characterized by rills and gullies at southwestern Taiwan

The Liji Badland near Taitung city is developed on a mélange of sediments including sandstone, siltstone and mudstone. Some of the materials are 'foreign' to the area and have transported from beyond the Taiwan area. This has been produced by the uplift of materials from the seabed, transport from the South Pacific Plate, debris flows and large deposition. This yields very diverse landforms after erosion (Fig. 3.18). Their rills and gullies evolved in a dendritic pattern. The rill development is interfered with the harder rock within the mudstone to make the forms more complicated.

3.3.3 Conglomerate Landscape

Conglomerates are a special type of sedimentary rocks, consisting of rounded pebbles of solid rock, sand, clay, carbonates or a mixture of all these. Thin layers of conglomerates are typical in areas with erratic and torrential rains, and some conglomerates resulting from stream deposition are several tens to hundreds of meters thick. Some of these conglomerate landscapes form unique badlands and weather to gravel (Fig. 3.19) (Chiang 2014; TCGS 2014b).

There are three conglomerate badland areas in Taiwan: Houyanshan, Chiuchiufong (which is literary translated into 'Ninety-nine Peaks') and Mt. Eighteen Buddhas. The naming of the badland hills refers to their shape or appearance. They reflect the youth of Taiwan's geology. In the conglomerate areas, gravels often pile up on riverbeds and river estuaries up to 300–400 hundred meters as a result of uplifting due to crustal compression since Quaternary. The gravels are poorly cemented, and thus susceptible to erosion.

Houyanshan is famous for its alluvial fans and widespread gullies. The gravels in Houyanshan are composed of quartz-arenite and quartz which are rounded by the effects of transportation. The top of Houyanshan is covered by 2–3 m of red soil. The red soil was formed as the overlying alluvial sandy sediments were weathered through high temperature and precipitation.

The conglomerates easily collapse and gravels pile up in the riverbeds and later moved downstream during heavy rains and typhoon seasons. Because of substantial erosion, one can observe sharp and steep edges on these conglomerate landforms. Between the gravels, the red soils work as a cement to maintain the steep slopes. Slopes located upstream with angles of 80–90° are characteristic of conglomerate

Fig. 3.16 A mildly outflowing mud volcano at Wushanding, Yenchao Geopark

Fig. 3.17 Mud volcano discharging methane that can be ignited, Yenchao Geopark

Fig. 3.18 Lichi mélange at Lichi Geopark

Fig. 3.19 Students observed and studied conglomerate rock at Liuguei, Maolin National Scenic Area

Fig. 3.20 Conglomerate rocks of Houyenshan Nature Reserve

landforms. Conglomerate gravels pile up alongside slopes and often form cliff cones. They may seem stable, but during heavy rains, they become hazardous if scoured by rivers (Fig. 3.20).

The gravel surface of Chiuchiufong collapsed during the 1999 Chi-Chi earthquake. The earthquakes loosened the gravels to allow further weathering. Unstable slopes and thin top soils make the area vegetation free. It is designated as one of Taiwan's first nature reserves. This designation aims to protect its unique conglomerate landforms. The Chi-Chi earthquake triggered a series of processes that destabilized the slopes and stripped the vegetation cover (Chiang et al. 2014).

Mt. Little Huang in Taitung, located in the Lichi geopark and to the south of East Rift Valley, represents a conglomeratic alluvial layer, created by erosion in the Beinan River catchment. Its steep slopes and intensive gullies indicate serious erosion. With tectonic compression, Mt. Little Huang has a different pattern of gullies from those of the rest eastern Taiwan (Fig. 3.21). Because of the tectonic processes, vertical joints evolved into gullies after heavy rainfall (Liu 2013).

Fig. 3.21 Conglomerate formation at Lichi Geopark

Fig. 3.22 Hot springs emitted at the tail of Turtle Island, Northeastern Coast

Fig. 3.23 Flat basaltic island at Penghu Geopark

3.3.4 Volcanic Landscapes

Due to Taiwan's geological location, considerable volcanic activities are to be recorded on the island. In addition to historical volcanic eruptions in northern Taiwan, basaltic magma eruptions are found within the Penghu archipelago. There are several different types of magma, each forming different plutonic rocks and volcanic landscapes (Fig. 3.22).

Magma and lava originating from inside the earth's crust are ejected at fracture zones creating volcanic landscapes. During and after eruptions, extrusive magma has formed cone-shaped rocks, bell-shaped rocks and even some flat topped rock formations. This is due to the different geo-chemical composition of materials that make up the magma. Lava made of andesite is viscous and mostly form cone-shaped volcanoes, such as the Chihsing Mountain in

Fig. 3.24 Fumaroles at Mt. Chihsing, Yangminshan National Park

Fig. 3.25 Giant basaltic columns at Hujin islet, a small islet in Penghu Marine Geopark

northern Taipei basin, whereas the basalt lava in the Penghu archipelago was not as viscous and formed mostly flat landscapes on cooling (Fig. 3.23).

Sulfur dioxide is released during volcanic eruptions and during cooling and crystallization. Thus pretty bright yellow sulfur crystals are formed. Even though most volcanic mountains are dormant in Taiwan, there are still sulfur holes and sulfur crystals. The composition and concentration of the minerals that make up the lava flow varies and its crystallization is variegated, forming crystals of beauty and diversity. The volcanic landscape areas of Taiwan often include geothermal resources, including hot springs, steam vents and pumps (Fig. 3.24).

3.3.5 Basaltic Landscape

The basalt in the Penghu archipelago was formed from lava that began to erupt 18 million years ago during the Miocene era. These eruptions ceased around 8 million years ago. During its 10 million years of eruptions, rocks of various formations and compositions intertwined with each other, resulting in the large diversity of outcrops on the archipelago today. Differential rates of cooling, extensions and contractions of basaltic rocks have resulted in a variety of joint patterns and structures that are in themselves art forms. Dinggoyu, Jishanyu and Hsiaobaisayu are three unique non-inhabited islets designated as Penghu Columnar Basalt Nature Reserve.

Basaltic mesas are made possible by the special characteristics of the basalt lavas of the Penghu archipelago. When

Fig. 3.26 The palaeo-soil layer at Penghu Marine Geopark

viewed from above and in cross-section, interlocking hexagonal patterns of basalt columns can be observed.

These basalt columns, when weathered, break-up into small and irregular rock blocks that accumulate at the bottom of the cliffs. Some palaeo-soil layers in between basaltic rocks have preserved the pre-existing land surface (Figs. 3.25 and 3.26).

The lotus platform (Fig. 3.27) is one of the famous attractions of the basaltic landscape of Tongpan Islet. An extrusion dome has been eroded by marine abrasion so that the interior is now revealed in the shore platform. This shows how softer materials or close spaced joint patterns have been eroded to form a concentric lotus-shape. Attractive as it is, tourists used to consider it a place for meditation because the lotus is a Buddhist symbol for tranquility and purity. The shallow hollow center of the dome becomes a tidal pool at low tide.

The inhabitants of Penghu used to construct walls, fences, shelters and houses with coral reefs and basalt rocks, as coral reefs were common materials in their daily life. These coral

Fig. 3.27 The lotus platform is a marine cut platform at Tongpan Islet

Fig. 3.28 Basaltic marine arch at Penghu Marine Geopark

Fig. 3.29 Marine cut platform at Penghu Geopark

reefs are also called "lo-go-ger" by the locals' dialect. The special structures constructed out of coral reefs have become local vernacular attractions, although many require maintenance and the maintenance cost is usually high if quality is desired.

Sea arches and sea cliffs, standing upright and forming the boundary between land and sea present spectacular views and are also constantly evolving. The bottoms of sea cliffs are under constant attack by the sea and suffer physical and chemical erosion leading to their eventual collapse (Fig. 3.28). When sea cliffs retreat, sometimes they collapsed and lowered to sea level and become shore platforms (Fig. 3.29). There are often tidal pools on the terraces that become cozy habitats for marine life.

3.3.6 Granite Landscapes

Granite landscapes are mainly found around the Kinmen and Matsu islets. Because of its hardness, many granite bedrocks

Fig. 3.30 Granite coast at the Matsu Geopark

were excavated for military tunnels during the cold war. The hardness of granite presented many challenges and hardship to those soldiers involved in the construction of military tunnels and defenses against Communist China. Today, the tunnels are considered valuable historical heritage and allow visitors the opportunity to reflect on Taiwan's military history and the cold war.

Granite landforms are often treeless due to the slow weathering rate of granite resulting in little or no soil formation. The beaches in Kinmen are composed of quartz sands making them crystal clear and under sunshine, sparkling. The granites in Matsu are often shaped into oceanic trenches and notches along the coast as the erosion processes exploit the joint systems. These processes also reveal various geological textures. Strong weathering provides Matsu with a variety of landscapes with beautiful rock joint patterns, which attract tourists.

Most Matsu islets are made of granite. With coastal processes and weathering, a catalogue of classic landforms has been formed. For example, precipitous sea cliffs, marine notches, marine arches, columns, caves and their distinctive and orderly joint patterns and unique textures are commonly found in Matsu islets (Figs. 3.30 and 3.31).

3.3.7 Glacial Landscapes

Because of the latitude and subtropical environment of Taiwan, glacial landscapes are solely found at the high mountains at around 3500 m above sea level. In the last ice age, the glacial processes left some erosion landscapes such as U-shaped troughs, possibly cirques and end moraines at the high mountains of Taiwan, such as Mt. Shei and Mt. Nanhuda (Lin 2012).

Whether glaciers occurred on Taiwan Island has been an historical academic debate for quite some time (Chyi 2003). The main question is whether the mountain ranges of Taiwan were high enough during the last ice age to ensure the formation of glaciers. Among other environmental conditions to make the existence of glaciers so doubtful are the effects of severe weathering and erosion that might have removed the evidence of glacial landforms. However, in the last ten years or so, more evidence for Taiwan's glacial landforms have been scientifically proven, such as striations on rocks. Earth scientists now commonly agree that glaciers once occurred on Taiwan (Fig. 3.32).

3.3.8 Coastal Landscapes

Taiwan is surrounded by the sea and the coastal zones are made up of a diversity of landscapes. Most coastal areas with particular and valuable landscapes are under the jurisdiction of the national Scenic Areas or National Park administrations. Northern Taiwan is famous for its rocky shores, beautiful sedimentary coasts are mainly found in western Taiwan, coral reef coasts are located in the south, and cliff coasts are famous in eastern Taiwan.

Some of the most spectacular coastal landforms occur on the north coast, where wintery northeastern monsoon and typhoons serve as significant weathering forces. Here beautifully colored rocks have been sculptured into an

Fig. 3.31 Domed granite landscape at Matsu Geopark

extraordinary kaleidoscope of shapes and colors. Geomorphologists consider weathering as the main process in making these subtle landscape differences.

Different rocks possess various resistances to erosion and weathering. Bays are formed where weak rocks are eroded; capes and headlands are formed where rocks resist erosion and weathering. Abrasion shore platforms and cuestas are special features along the northern coast. Chess board (Tofu) rocks, scattered along the coast are spectacular and are the results of weathering and erosion processes combined with its jointed rock formation (Fig. 3.33).

Taiwan is also noted for the occurrence of marine terraces. These are shore platforms lifted out of the sea by tectonic processes or exposed due to the fall of sea level. Cliff coasts are the best places for identifying the different strength of various layers of sedimentation due to its varied

Fig. 3.32 The highest point of Mt. Shei has some indicators of glacier activities

Fig. 3.33 Wave cut chess board rocks at Yehliu Geopark

Fig. 3.34 Coastal cliffs at northeastern Taiwan

resistance to erosion and weathering (Fig. 3.34). That is to say, the shape of cliffs is often controlled by the underlying sedimentary sequence of rocks of different strength. Taiwan has a remarkable sequence of dated surfaces especially on the east and south coasts where it has been possible to estimate the rate of uplift of the Coastal Ranges.

At the sandy beaches of the coasts, bio-diversity is very high. In particular in the northern and south-wester river estuaries, natural reserves marine lives are particular impressive. Settlements often attract urbanites (Fig. 3.35). For instance, mangroves in the estuary of Tsenwen River at south-western Taiwan are very valuable and rare ecosystems which form important wild life reserves (Fig. 3.36). In the southwestern coasts, together with the development of unique spits the enclosed sediments reveal the history of shoreline and river evolution. Oyster farms at the south-western wetlands, along with its excellent habitate for annual migrant birds, are particularly impressive as tourist

Fig. 3.35 Coastal village at Chinkunsen, Southwestern coast of Taiwan

Fig. 3.36 Mangrove at the estuary of Taishue River, Northern Taiwan

attraction. The migrant bird habitate makes the surrounding communities even eager to take car of the resourceful and clean environment, which is beneficial for good harvest for the aqua-farmers as well. As all sorts of fish and shell fish farming is a vital economy in the southwestern coastal settlements, their success depends on strong sustainable coastal management. This is as significant because the areas are subject to flooding at the summer and fall seasons.

References

Camanni, Giovanni, et al. (2014). Basin inversion in Central Taiwan and its importance for seismic hazard. *Geology, 42*(2), 147–150.

Chyi, S. J. (2006). Riverine evolution and potential hazards. *Journal of Geographical Science, 19,* 51–70. (in Chinese).

Chyi, S. J. (2002). *The conservation guidebook of Tainan's geomorphological landscapes*. Kaohsiung, Taiwan: National Kaohsiung Normal University. (in Chinese).

Chyi, S. J. (2003). The tear drop of the heavenly god—Is Chia-min Lake a giant impact crater or a glacial cirque? *Journal of the Geographical Society of Taiwan, 32,* 1–16. (in Chinese).

Chiang, T. R. (Ed.). (2014). *The great geological exploration of Central Taiwan*. New Taipei City, Taiwan: Taiwan's Central Geological Survey. (in Chinese).

Lin, J. C., & Su, S. J. (2013a). *The dynamic landscapes of Taiwan*. Taipei, Taiwan: National Taiwan University. (in Chinese).

Lin, J. C., & Su, S. J. (2013b). *Landscape conservation of Taiwan*. Taipei, Taiwan: National Taiwan University. (in Chinese).

Lin, T. C. (2012). *The great geological exploration of Taiwan*. New Taipei City, Taiwan: Taiwan's Central Geological Survey. (in Chinese).

Liu, Y. S. (2013). *The environmental education guidebook for Li-Chi Badland Geopark*. Hualien, Taiwan: National Tunghua University. (in Chinese).

Taiwan Central Geological Survey. (2014a). Geologic time is now! *Geology, 33*(1), 5. (in Chinese).

Taiwan Central Geological Survey. (2014b). Defining the spatial boundary of sensitive geological areas and its policy implication. *Geology, 33*(1), 12–13. (in Chinese).

4 Geoparks Manifesting Human Relationships to the Environment

4.1 Socio-cultural Environment of Geoparks

The socio-cultural environment of Taiwan's geoparks and their surroundings have been shaped by history and the human activities taking place under each geopark's unique geological and geomorphological conditions (Su et al. 2014). Human activities have had a great influence on the natural environment and the landscapes with socio-economic implication (Goudie 2018; Meyer 1996). The livelihood, daily activities and routines, religious activities and beliefs of the geopark community have been significant in the relationship of the local people and their environment. Because the natural environment of each geopark is different, each geopark and its community has a unique face of socio-cultural landscape.

The materials, techniques and technologies which determine the life supporting, social, political and economic activities of each geopark need to be identified because these human activities are adaptive with respect to the natural environment. That is to say, each geopark community's socio-economic landscape complex is a result of the interaction of human beings and their environment.

A great example of the human-environment interactions come to mind instantly is the underground tunnels in Matsu Geopark. Matsu islets have the most densely distributed underground tunnels in the world due to the national defense need during the cold war between Taiwan and China. Matsu islets are Taiwan's closest frontier to China, only 9.5 km away at the closest point to Chinese territory. During the cold war period the Taiwan government prepared Matsu as a defense base against Communist China's invasion by building an underground tunnel network (Fig. 4.1). As the islets are composed of mainly granite rocks, the tunnels could become a solid safe haven when the islets were militarily attacked by China. However, the hardness and rigidity of the granite rocks became great hurdles for the military personnel and soldiers who were assigned the task. There are stories told by the soldiers as well as the locals about the hardship of the tunnel making (Ley et al. 2017). Today, when we enjoy the beautiful and memorable tunnels on Matsu islets, and even when we taste the appetizing liquors that were produced using the tunnels as the best wine cellars, we will certainly appreciate and understand the tunnels as a socio-cultural landscape of the military confrontational time (Fig. 4.2) (Su 2017).

Socio-cultural landscapes have an important historical significance. Taking advantage of the past imprints of human culture and activities in the natural environment, geoparks and their communities are able to learn from the nature, the past human activities and environmental history. With the learning and realization of geopark experience, a sound human-nature relationship can possibly be understood well. As the promotion of Taiwan geoparks has combined scientific knowledge, local knowledge and local actions, how the human society interacts with the natural and artificial environments has been a critical source of community awareness and enlightenment.

Geopark management and managerial practice is another aspect that is not only knowledge-based, but also bound by local identity and culture, skills and technique, and the local environment and knowledge of the changing nature. All should be about lived cultures and performed daily practices. Even the concept and practice of "environment management" should be replaced with "environment governance" in which the actors of society determine and act on goals and priorities related to natural resources (IUCN 2017). As the concept of governance has been adopted by many global institutes, the long debated term of "governance" in the academics has come to some typical aspects of making it extended and including more inclusive perspectives.

For example, Lemos et al. (2008) argued against the sole consideration of the state and/or market actors. They reviewed four domains of governance, including globalization, decentralization, market and individual incentives-based governance, and cross-scale governance, and argued for the emerging hybrid modes of governance across the state-market-community divisions. For asking questions of environmental change and governance, such divisions and

J.-C. Lin and S.-J. Su, *Geoparks of Taiwan*, Geoheritage, Geoparks and Geotourism,
https://doi.org/10.1007/978-3-030-04895-2_4

Fig. 4.1 A military tunnel drilled by the soldiers during the cold war period is now turned in a tourist attraction of Matsu

Fig. 4.2 Rice wine kept at the tunnel which was considered the best place for keeping and chilling wine

coupled social and natural systems may in fact beneficial for a bottom-up approach of geopark development. As Armitage et al. (2012) have also argued simply, new actors play critical roles in decision-making, and new mechanisms and forums for decision making take up increased significance for conservation. Practices of conservation then have to consider institutional fit, institutional scale, adaptiveness, flexibility, learning, knowledge coproduction from diverse sources, emerging new actors and their roles, and changing expectations through time and legitimacy change. In comparing with the experience of promoting and developing geoparks, these considerations are all but significant and critical to make it a success.

One critical perspective about good governance of the geopark development is the historical context and the contextual depth of the place and community. The idea of Japanese scholar Sasaki Tsunen (Sasaki 1997) about landscape, scenery and terroir is an excellent way of emphasizing the significance of rooted local context through time, culture and space. It is not only about the local economy, but more importantly and more fundamentally about a local socio-economy that consequently reinforce the local values and capacities to be able to engage in conversation with the global. The salt fields of the Southwestern Wetland Geopark are a good example. Located in the sub-tropical zone, the southwestern wetland coasts of Taiwan were excellently suited for salt production. Flat coastal fields, ample sunshine, abundant local labor supply with skills, and a huge demand of both the local market and the then Qing dynasty's market were all pillars to drive continuous salt production on this relatively limited area of southwestern Taiwan since the Qing dynasty (Fig. 4.3). Salt was the material that linked the Qing dynasty with the now island state of Taiwan. In addition, salt production of the southwestern wetland Taiwan created a particular way of life, unique social cultures, wretchedness literature of life, and even endemic diseases in the local context.

Today, salt fields as a culture landscape also provide the best sites for environmental education which introduces the public to the proper use of sea water for human needs and to the salt workers' hard life, wisdom and technology. Unfortunately, the salt production industry of the southwestern area no longer exists and the salt landscape is only kept intact for environmental educational and leisure purposes. The historical socio-cultural landscape of the salt fields however, remains an important educational site.

Socio-cultural landscapes in geoparks are based on its geology, geomorphology, history and related environmental conditions, but are often shaped and altered by the needs and desire of human society (Wang 2017). Therefore, geoparks are a showcase of human history and activities in their natural state. The socio-cultural landscape and environment are important because they are relics for the present generation and which help future generations to understand their own environment. The understanding of the living environment allows people to know their own cultures in the presence of nature. Such understanding of the living environment will in turn generate the power of appreciation and identity, which is important for safe guarding the environment. This is exactly the essence of local conservation and protection of the environment in the context of geoparks.

Geoparks in Taiwan have another important public meaning. Many Taiwanese geoparks are located within places with special scenic or esthetic value (Lin 2014). Some are administrated and maintained partly under National Scenic Administration and some within Nature Reserve legislation. Therefore, the communities surrounding the special and valuable geological environments have experienced many visitors and tourism activities. For the geopark communities under National Scenic Administration, various interpretation practices have been conducted but without much geological and environmental content. This is a window of educational opportunity for the geopark communities to learn about their own environment and to become interpreters for their own geopark. The Yehliu geopark is a case where local business men and women, after a long history of doing business for tourists, came to realize that there were more layers of meaning in their environment (Shen et al. Shen 2017a). With coastal erosion and weathering, the famous Queens's Head monument could be gone forever in

Fig. 4.3 A panoramic view of abandoned salt fields at the Southwest Wetland Geopark

one incident. This is the knowledge that prompts the locals to begin to value the local landscapes even further and become protectors of their landscape.

For the geoparks which contain Nature Reserve sites, a long history of careful implementation of conservation has trained the locals to be aware of the significance of the nature (e.g., Lin 2015; Chiu 2017). Here, promoting geoparks becomes a much easier task. The locals understand the essence of geoparks easily as they proceed to join the geopark community of Taiwan. They also become the vanguards and good practitioners for geopark promotion. Such geopark communities are often able to care for the sites of special value, and are often able to become initiators for the better practice of conservation. Such characteristics certainly come from the daily routine of caring of the nature. In these communities the people who are close to the nature and watch it with care may be farmers or people relying for their livelihood on the land and nature. The land and nature are what support their life, and the care becomes authentic. There are, however, some people who care for the nature out of the love and appreciation of it. There are people whose childhood memories are attached to the geopark area. The Yenchao bad land and mud volcano Geopark is the perfect case where the extreme landscapes are also the loved home land to many people (Figs. 4.4 and 4.5).

Therefore, the practice of the National Scenic Area and Nature Reserve managements related to the geopark promotion often seeks to provide the necessary tools and instruments to facilitate environmental training and education for the locals. These supports, along with other planning instruments and conservation tools and skills, serve as critical utilities for the local people to be effective in geopark development and management for sustainable future (Wang et al. 1999; Wang et al. 2002; Lee et al. 2010).

Geo-diversity is not only manifest in nature, but also in human culture (Ho 2009; Lee 2012). Cities and towns, hills and plains, mountains and coasts all bear the hallmarks of human activity, where the natural landscape has been modified to meet the food, shelter and communication needs of humankind. Buildings, temples, and settlements are where we may witness the interactions of human society with the natural environment. Economic activities may produce landscapes where one can observe how human societies interact with nature, utilize nature, and produce culturally distinctive landscapes (Figs. 4.6, 4.7, 4.8 and 4.9). Adapting to local milieu, such as certain architectural styles to suit natural environment, and using local material, mostly heavy granites, are critical elements of the local architectures and cultures. For instance, typical stone buildings at Chinbi village of the Matsu Geopark act as a link between the geological environment and the traditional community adaptions to the environment. The degree of elaboration of temple architectures in different places may illustrate varied local environment and architectural cultures (Ley et al. 2017).

Inappropriate developments in environmentally sensitive areas often trigger hazards beyond our control (Chen 2017; Lee 2014; Lin 2011). Steep slopes, tidal flats, riverbeds, and areas subjected to subsidence are typical examples of sites that are unsuitable for human activity or development. Such developments not only bring harmful results to the

Fig. 4.4 A green canopy leads to a mud volcano of the Yenchao Geopark

Fig. 4.5 A local interpreter guides the visitors at the Yenchao Geopark

Fig. 4.6 A form of coastal protection with concrete armor blocks at northern Taiwan is considered a poor design today

environment and ecology, but also to human settlements, costing lives and damaging individual properties. Education at geopark often helps sensible and prospective decision making.

The surveys and audits undertaken as part of the landscape conservation process also provide a useful indicator of environmental quality. For the past decades, Taiwan has achieved significant economic growth at great expense to the environment. It is now the appropriate time for the central and local governments, government agencies, schools, communities and individuals to subscribe to the landscape conservation agenda and respect and act in synergy with nature. With this mindset, Taiwan will be able to achieve sustainability facing rapid global environmental change.

Fig. 4.7 Elaborately carved beams and painted rafters of the temple architecture at Penghu Geopark representing religious significance to the fishermen

Fig. 4.8 A temple at Matsu Geopark with simplified decors than the one in Penghu indicating very different architectural and material cultures

4.2 Community Building Projects as Catalysts for Geoparks

Due to the nationalistic movement for asserting and consolidating national cultural and knowledge influence in the 1970s, the then-KMT controlled government launched an academic programme to survey and establish a knowledge base for Taiwan policy. The programme was entitled "Zho-Da" program, which merges the name of Zho-shui River and Da-du River. The program aimed mainly to investigate the environment of the two river catchments with respect to archaeology and ecology through an interdisciplinary approach. It was a time when the country of Taiwan experienced political and international hardship due to its withdrawal from the UN when the People's Republic of China (PRC) became the sole representative for East Asian Chinese cultures. Having no intention to reveal the details for the country's political hardship and predicament, it was often argued that the 1990s community building programs in

Fig. 4.9 The top of a traditional stone house representing a seal with official significance, Matsu Geopark

Taiwan were to encourage an authentic Taiwan's identity. The viewpoint that Taiwan wished to recover China and to take over the PRC government was considered politically naive and thus abandoned.

The Zho-Da program, as a national academic endeavor to establish Taiwan-based knowledge of Taiwan environment and cultures, was quite a superb academic and political move. It was followed later by a national movement for community building for the in situ sense of identity which is largely considered a movement for Taiwan nation building today by analysts. There are a few principles of the community building program that turn out to be the accidental foundation to enable geopark promotion to be implemented with ease:

- To build a new society through national programs of education, science and good hard work.
- To encourage community solidarity. This is now part of the rationale for the success of geoparks.
- Such community solidarity is not given and should not be taken for granted as it is a result of long term and painstaking processes, supported by the wishes of people for identity building of Taiwan. It is part of the national program of community building.

A central belief was that through the investigation of landscape, ecology, culture and industry, the environmental resources of the country could be identified and cherished. The communities thus can become aware of the pros and cons of following an eco-friendly way of conservation and able to decide the appropriate and suitable plans for their environments. With enough and proper knowledge of their own environment and community, self-reliance and self-improvement can be sustainable and helpful for the whole community. With further networking among all the communities, both adjacent and faraway places can benefit from the shared learning and knowledge.

Through environmental education and good practice of environmental management, further solid concepts of conservation and the good practice of conservation can be achieved. With the good training for environmental interpretation and rounded understanding of environment, the local people are able to represent their milieu and to connect themselves to the neighboring communities.

The cooperation among all stakeholders either within or outside of the community is one critical element. Cohesion and leadership inside the community, which required participation, sharing and organization, are the most significant. Good leadership results in cohesion which provides platforms for working together. With the mentality of self-improvement of the community, capacity building by education, study and training becomes a creative pursuit among most communities. This is one window opportunity where the academia and the public sectors are sought for certain expertise and knowledge. With the contribution of academia, most geopark communities can work with confidence and see it as a supportive network. Other sources of supports may be from NGOs as sponsors, environmental professionals, policy tools, laws and regulations. Sometimes strategic alliances among diverse stakeholders may prove to be helpful.

During implementation of geopark promotion plans, stakeholders are often asked to participate in various ways according to community development strategies and the pursuit of certain elements for sustainability, either of environment, economy or/and society. Economically, the geopark promotion is often geared toward the geo-products that enable the local agriculture or any land-borne economy to have chances of becoming value-added. In terms of societal value, geopark promotion include the improvement of people's living conditions, wellbeing, improved school education and creating employment opportunities to attract young people to return to their own communities. It will eventually reduce regional disparity by mending the disparity of human resources. In the analysis of community building for conservation, the geopark promotion can have at least two principles to consider:

- To identify the national landscape conservation and its policy as the way to mobilize social bottom-up power. This is important because Taiwan's environment relies upon 23 million people who will take care of it, and also because Taiwan is part of the global environmental system and has a responsibility to look after it for all people.
- To engage in or stimulate dialogues between the knowledge bodies of geology, geomorphology and the living nature, and Taiwan's society to create a rounded and transdisciplinary platform for better conservation practice. With the values of geoparks as the fundamental concept, it is possible to ensure a bottom-up endeavor for sustainable environmental future for Taiwan.

4.3 Concluding Remarks

The planet earth is the only living sphere that we have and depend on for civilization. With more extreme changes in the environment and more detrimental hazardous events occur in every corner of the world, we come to realize that only through a careful scrutiny of the history of human-environmental relationship, the human society can be repositioning its role for a sustainable earth. The interaction of human life with the nature then become a fundamental approach of understanding our situation. The geopark approach is one with pragmatically sustainable purpose, creative tools and human agency and engagement.

The diversity of the dynamic landscape of Taiwan makes it a beautiful island. Nature has its own rhythm in making Taiwan what it is. However, in utilizing and co-existing with the nature for human comfort, innovation and civilization, we are adding tremendous pressure on the natural environment. Therefore, for landscape conservation that could be of utility for protecting our only planet, it requires environmental education. With the understanding and appreciation of how the human societies have evolved through time and space in making the planet earth what it is now, we become more careful about what we value and what we should value. Therefore, geopark practices become a window of opportunity to a way of life that may promote sustainability for human existence and for leisure life, eco-tourism.

Landscape conservation educates and delivers the knowledge-base and skills and techniques that contribute to hazard mitigation and containment. Better yet, with proper and sustainable planning of good knowledge ahead of time, societies will know what, how and when to act or not to act in order to conserve our environment. The task of landscape conservation requires public participation. For this reason, this book is aimed at introducing the various landscapes of Taiwan and its many cases and aspects of work and implementation to the public participating in building a sustainable future through geopark promotion.

References

Armitage, D., de Loë, R., & Plummer, R. (2012). Environmental governance and its implications for conservation practice. *Conservation Letters: A Journal of the Society of Conservation Biology, 5* (4), 245–255.

Chen, Y. H. (2017). A geopark filled with life and death stories. In *Taiwan on geological scale: Tracing Taiwan geology from geoparks* (pp. 148–171). Taipei, Taiwan: Metropolis Publisher. (in Chinese).

Chiu, Y. Y. (2017) A Muddy Geopark with sweet fruit. In *Taiwan on geological scale: Tracing Taiwan geology from geoparks* (pp. 172–209). Taipei, Taiwan: Metropolis Publisher. (in Chinese).

Goudie, A. S. (2018). *The human impact on the natural environment: Past, present and future* (8th ed.). New York: Wiley-Blackwell.

Ho, L. T. (2009). Landscape diversity and its conservation. *Scientific Development, 439,* 22–29. (in Chinese).

IUCN. (2017). IUCN webpage on environmental law for sustainability. https://www.iucn.org/theme/environmental-law/our-work/governance-and-measAccessed August 2017.

Lemos, M. C., & Agrawal, A. (2008). Environmental governance. *Annual Review of Environment and Resources, 31*(1), 297–325.

Lee, C. T. (2012). *The Geo-diversity of Tsaoling*. Taipei: Forestry Bureau of Taiwan and National Taiwan University co-published. (in Chinese).

Lee, C. T. (2014). Landslides and Tsaoling Geopark. *Geology, 33*(1), 31–35.

Lee, K. C., Wang, S., Ho, L. T., & Chang, S. C. (2010). Community-based framework for planning geoparks. *Taiwan Forestry, 36*(5), 13–21. (in Chinese).

Ley, S. Y, Chuang, R. L., & Wang, W. (2017). An Epoch of Geopark of iron and tears. In *Taiwan on Geological Scale: tracing Taiwan geology from geoparks* (pp. 308–355). Taipei, Taiwan: Metropolis Publisher. (in Chinese).

Lin, J. C. (2011). *Classroom of Taiwan geomorphology*. Taipei, Taiwan: The Owl Publisher. (in Chinese).

Lin, J. C. (2014). Promoting Taiwan Geopark network. *Geology, 33*(1), 20–22. (in Chinese).

Lin, J. C. (2015). *The inventory of and interpretation of landscape resources of Hu-Jin and Tongpan islets*. Penghu, Taiwan: National Scenic Area Administration of Penghu. (in Chinese).

Meyer, W. B. (1996). *Human impact on the earth*. Cambridge, UK: Cambridge University Press.

Sasaki, T. (1997). *Ten years for landscape, hundred years for scenery and thousand years for terroir—Legacies for the 21st century*. Tokyo, Japan: Soyo-sha Corporation. (in Japanese).

Su, S. J., & Lin, J. C. (2014). The environmental and social implication of geoparks. *Geology, 33*(1), 56–69. (in Chinese).

Su, S. J. (2017). Matsu's historical development and its implication for geopark development. *Geology, 36*(4), 55–62. (in Chinese).

Shen, S. T., & Lin, S.F. (2017a) Queen's head and her coastal geopark. In *Taiwan on geological scale: Tracing Taiwan geology from geoparks* (pp. 96–127). Taipei, Taiwan: Metropolis Publisher. (in Chinese).

Wang, C. Y. (2017). The updated regulation for the cultural heritage of natural landscape and monuments. *Communication for Landscape Conservation, 45,* 2–7. (in Chinese).

Wang, S., & Ho, L. T. (1999). Introduction to landscape conservation techniques. *Communication for Landscape Conservation, 11,* 6–8. (in Chinese).

Wang, S., & Lee, K. C. (2002). *Report for establishing geoparks and its environmental monitoring and management*. Taipei, Taiwan: Council for Agriculture Taiwan. (in Chinese).

5 Environmental Education for Geoparks—Practices and Challenges

5.1 Structural Support for Geoparks

In June 2010, Taiwan launched the national *Environmental Education Act* after two decades of debates and general discussion. As the Act is the 5th national environmental education law in the world, Taiwan's society celebrated the potential environmental effects of the Act along with a series of educational reforms (Lin 2013; Yeh 2017). In November of 2017, however, a revision of the law was initiated to adjust to the rapidly changing society and environment of Taiwan. Among the revisions, the change in Article 1 is particularly important that focused on national environmental ethics and responsibility. Article 1 was revised to additionally include the significance of understanding the environment. The reason for this stemmed from the feedback on educational implementation. Proper implementation of environmental education has to fit local needs, varying contents and, thus, methods had to be formulated in order to respect that ecology, lives, and social contexts vary across the small island state. Needless to say, the goal of the *Environmental Education Act* is to maintain the ecological balance of the environment, to respect life, and to foster social and environmental justice. All these are convened to promote sustainable development, with environmentally aware citizens, sound environmentally friendly communities and rigorous environmental studies. For the citizens, communities and studies, such environmental laws have coincided with the efforts to establish and facilitate geoparks.

This set the stage to facilitate the future of Taiwan's environment through the promotion of geoparks. Since one of the missions of geoparks is to develop environmental education through relevant activities, environmental regulations and lawful facilities, the *Environmental Education Act* provides principles, policies and powers and it specifies the responsibilities of agencies. It promotes, gives incentives and even enforce penalties. For the geoparks of Taiwan, The *Environmental Educational Act* comes as a facilitator and benefits their promotion through its collaboration with schools and communities.

This important advance is enhanced even further by the *Culture Heritage Act* that was successfully revised with the addition of 2016 that include natural landscapes/sites and geoparks as part of Taiwan's heritage. Geoparks, natural heritages and natural monuments with special, scientific and educational values are specifically mentioned in the act. Previously, amendments to the *Cultural Heritage Act* had been discussed for over ten years and progress was relatively slow (Taiwan Geomorphology Study Group 2013). As the progress on prioritizing landscape and natural conservation was disappointing, societal attention and pressure accrued. Various NGO groups demanded amendments to the heritage act because they were against land enclosure, and wished to eradicate all possibility of developers from dismantling historical heritage sites. The societal spirit of looking after natural sites/landscapes has been high. It came as a happy surprise as the academia, communities and NGO groups welcomed the new addition of the heritage act to incorporate natural heritages and geoparks into the law. The newly amended *Cultural Heritage Act* of 2016 becomes a powerful vehicle for promoting geoparks and natural landscape conservation.

The Culture Heritage Act sought to raise the awareness on cultural and natural landscape conservation through education and daily life practice. The development of tourism and local participation became one significant means for such conservation. Tourism and local or stakeholder participation both succeeded in drawing the public attention to guarding the sites of interest against demolition or damage. This function of incorporating conservation into legislation is in accord with the practice of geo-tourism and eco-tourism in geoparks.

Apart from these structural supports for geoparks, there are numerous other avenues and methods that enhance the imagination even more for environmental education and cultural heritage conservation. In practice, there are many learning opportunities and ways of improvement in realizing lawful regulations. Thus, it is fair to say the practice of conservation for geoparks has been diverse and creative in

J.-C. Lin and S.-J. Su, *Geoparks of Taiwan*, Geoheritage, Geoparks and Geotourism,
https://doi.org/10.1007/978-3-030-04895-2_5

Taiwan. For instance, the *Environmental Education Act* fosters a legitimate need for environmental education. By law all government workers and school employers should take 4 h of environmental education training courses annually, either by attending lectures, or visiting national parks, geoparks, national scenic areas, or other designated places of environmental education. Such requirement and practice provides all environmental educational sites, including geoparks, the chances of formulating and improving courses for environmental education. It is certainly one significant avenue for geoparks and their communities to become focused on providing appropriate services. By generating synergies between various laws and civilian forces, the missions of geoparks become more diverse and imaginative. Creative activities and educational methods are in demand to promote the chance for geoparks to be more effective for conserving natural landscape (Chiu 2017; Chuang and Wang 2017; Ley et al. 2017; Shen and Lin 2017; Su and Lin, 2014).

Today, some geoparks of Taiwan have concrete environmental educational programs that can be utilized by the local schools and communities and even by visitors. Some, however, are still trying to create relevant educational programs for better practice. By the way, late comers are not necessarily less mature, they could benefit from the experience of others through the well-developed networking system of the geoparks.

The four main core values of geoparks, landscape conservation, environmental education, geo-tourism and local participation, are not only guiding polestars but also inspiring directions for geopark communities. Most geopark communities, through long term networking and learning from each other and with the help of scholarly assistance from University communities, have been highly motivated for creating interpretation programs how to best guide visitors and their own local youth. It is mostly due to this development that the societal and educational tools can fulfill the concept of environmental education. Through the introduction of local environment to park users, through multi-media, guided tours and pragmatically designed geo-products (e.g., the so-called DIY programs), local communities enjoy the fun and sense of achievement in making themselves known and understood by visitors. This is a measure of the success of the policy.

5.2 Environmental Education Practice in Geoparks

Geoparks can pass on to the public the conservation concepts of environmental knowledge and environmental protection through varieties of educational opportunities and activities. Living museums, interpretive and educational centers, guided geo-tours/eco-tours, popular literature and maps, modern media publicity etc. are just some examples of a geopark's practice for environmental education. Through the conduction of research on the local environment and through education programs provided by independent community organizations and relevant departments of universities, residents and visitors acquaint themselves with their local environment.

The people who take care of their own living environment, are usually people who do it out of their love of the place, *topophilia*. Often no pragmatic material desire or monetary return is expected. Therefore, the care of environment becomes a must 'do' and mandate. For instance, people who love and enjoy open water diving at the north-eastern coast of Taiwan are certainly concerned when there is an oil spill on the northern coast (Shen and Lin 2017). Under the sea, plenty of opportunity for exploring marine environment and sea life exist. In such places, nature speaks for itself. No extra persuasion is needed for the people who love and enjoy open water (Figs. 5.1, 5.2 and 5.3). With good care of the marine environment, both novice and professional divers will enjoy the sea world. Coastal geoparks provide opportunities for people, either local or non-local, to get closer to nature and to understand it with the assistance of local interpretation facilities. More than often the sites at geoparks themselves could become the facilities and venues for environmental education and enjoyment (Wu and Chang 2017, 2018; Wu and Wang 2012).

Geoparks aim to conserve both natural and cultural landscapes not only by preventing damage, but also by making them educationally and academically valuable sites. A variety of activities and opportunities, such as excursions for school groups, seminars and conservation activities, take place to introduce local landscape and environment to residents, teachers, students, and tourists (Wu and Chang, 2017, 2018). This means that geoparks serve as a window for environmental education. Understanding the natural and cultural environment is therefore a pragmatic starting point for conservation in geoparks. It in turn usually reinforces the local consciousness, pride in place and cultural identity. These are beneficial to environmental and landscape conservation and attractive to visitors. When tourists have deep understanding of the places they visit and seize the spirit of the love of the locals for their own place, visitors gain greater appreciation and learn to cherish the place for themselves.

Geoparks have another function in Taiwan's school curriculum. During the school curriculum reforms of the early 2000s, the national educational committee for geography as a school subject added the concept of "landscape conservation". That turned out to be a critical turning point in environmental conservation and for the foundation of geoparks. It may be argued that the changes to the school

Fig. 5.1 Scuba diving training at Northeast Coastal Bitou-Longdon Geopark

Fig. 5.2 Divers enjoing the marine life at Bitou-Longdon Geopark

curriculum was a productive foundation for the establishment of the *Environmental Education Act* and a versatile groundwork for geoparks.

Geoparks supported by the local people provide many aspects of environmental knowledge and information for visitors who are enriched by the local support. Friendly and personally experienced environmental knowledge, plant seeds of knowledge for conservation in the minds of visitors (Chiu 2017). For this to be realized, it takes very hard work. For example, some environmental educationists at geoparks are good at guiding tourists and through their experience the concepts are easily conveyed and produce peak experience. Guiding visitors is necessary to successfully navigate the bad land landforms in order for visitors to understand the texture,

Fig. 5.3 Snorkeling course at Bitou-Londong Geopark

Fig. 5.4 Guided observation yields greater personal understanding in the Yenchao Geopark

slope and fragileness of the muddy bad land (Figs. 5.4 and 5.5).

Another example is to show and appreciate the magnitude of the height and dimension of basaltic columns by comparing them to their own bodies. The local interpreters guide visitors to use their own bodies as a scalar measure to pinpoint the height of Penghu Marine Geopark's famous basaltic columns. This makes it easier for visitors to realize the extent of the volcanic processes and the complexity of magma crystallization (Figs. 5.6 and 5.7).

Geopark communities have become the main drivers for interpretation activities due to their intimate understanding and appreciation of their own places as natural laboratories (Chuang and Wang 2017). They know best the potentials of their geoparks and are in the position to demonstrate the interaction between human society and the environment. To prepare themselves to become good interpreters, the local communities of geoparks go through a process of learning and networking. Sometimes scientific information through training programs has to be provided on issues like water

Fig. 5.5 Navigating the muddy bad land enhances the understanding of landform and erosion at Yenchao Geopark

Fig. 5.6 Visitors standing in front of the basaltic columns of Penghu Marine Geopark

quality of the wetland or the seasonal rhythm of historical salt fields (Fig. 5.8). Some communities have even devoted their own property to become geopark hotels and provide hotel rooms free of charge for geopark activities and environmental education (Fig. 5.9) (Chen 2017). These all provide excellent motivating chances for the public to understand and cherish the environment. With the facilitating support and the implementation of the national *Environmental Education* and *Cultural Heritage Acts*, these developments have become successful.

5.3 Geopark Practices of Environmental Education

By attracting visitors to geo-sites or geoparks to experience their particular beauty, geo-tourism activities fulfil the local economic needs by empowering the local community with local knowledge and interpretation skills. Such knowledge-oriented capacity development enhances the local economy and, thus, provides a better and sustainable

Fig. 5.7 The diameter of this basaltic column is about 1.8 m at Penghu Marine Geopark

Fig. 5.8 Getting to know the science of their "own place" by learning how to monitor water quality at Southwest Coastal Wetland Geopark

future. This also encourages local participation in environmental protection and conservation activities, and creates value-added sustainable geo-tourism. Such local mechanism is important because a bottom-up process is triggering the developmental momentum of a region (Figs. 5.10 and 5.11).

5.3.1 The Case of Yehliu Geopark

The most typical case of environmental education in the geoparks of Taiwan is the Yehliu Geopark. The Yehliu geopark is one of the first scenic areas obtaining the certificate for environmental education by the Environmental Protection Agency of Taiwan. It develops and provides 14 environmental courses for visitors of different ages and needs. The geopark's Nature Center, which used to be an army camp before it was renovated, is now a sparkling site for environmental education. At this center local school children and young visitors learn about coastal landscapes (Fig. 5.12). Every year the Yehliu Geopark provides an annual budget of US$200,000 devoted to environmental education for schools and organizations around northern Taiwan.

Fig. 5.9 The hotel turned into an Environmental Education Center at Tsaoling Geopark

Fig. 5.10 Environmental Education Center at Taijan National Park

The Nature Center has many functions. The place is not only used as an indoor classroom but also as an exhibition hall for coastal landscape illustrations and local cultural artifacts. Both students' and adults' environmental courses is supported by various relevant educational materials. The center also provides training on inspecting and managing marine life, as this is extremely important for people living in a small island state.

5.3.2 The Case of the Yenchao Mud Volcano and Badland Geopark

Yenchao Geopark is famous for its mud-volcanoes, rugged gullies of mudstone and flammable gas outlets. It is called "bad land" due to the mud landscape that does not easily bear agriculture or vegetation. It has been designed as a Nature Reserve under the *Cultural Heritage Act* for more

Fig. 5.11 Students at a field study on Lichi mélange formation, Lichi Geopark

Fig. 5.12 Students studying mushroom rocks at Yehliu Geopark

than a decade. Visitors can only visit the sites with a guided tour provided by the local organized interpreters helping visitors to appreciate and understand the unique landscape and local ecology. That lays the foundation for the geopark's environmental education program combined with activities for visitors (Fig. 5.13).

Close to the Chi-shan Fault, underground natural gases of methane and ethane usually escape through ground joints or passage ways that provides occasionally water and other fluid flows and unobstructed underground passage to the surface. Under certain conditions, the gas escaping creates bubbles on top of the surface that attract visitors to observe the natural wonder; some even light the top of the mud volcanos with fire to make a splendid scene (Figs. 5.14 and 5.15). As scientific studies try to record the eruption behavior of the mud volcanos, guides make their own

Fig. 5.13 Guided interpretation before entering the nature reserve at Yenchao Geopark

observations of the intervals and seasonal rhythm, and develop, thus their own theories and understanding.

As the rill and gully erosion of the mudstone area develops under the influence of extreme weather phenomena, in particular extreme rainfall or typhoon rains, land managers find it challenging to maintain the area. The local farmers, who used to know the place and control their production activity well, have started to realize the new challenges caused by climate change. Water irrigation and mud flows become of concern. This certainly brings the need for

Fig. 5.14 The natural gas at a mud pit can be lit to become a scene for environmental education at Yenchao Geopark

Fig. 5.15 A mud volcanos in Yenchao Geopark

serious research in order to find a balance between agricultural land use and the natural erosion. Through continuing observation and environmental education, managers and farmers are becoming aware of what needs to be closely observed.

5.3.3 The Case of Matsu Geopark

School exchange programs, initiated by the Matsu Geopark with the support of the scenic area authority, are one of the best practices for environmental education within a geopark. The Matsu National Scenic Area authority annually funds at least 60 primary school students to participate in the geopark exchange activities. Even with a limited budget, the students of the Matsu islets are able to visit the geoparks of Taiwan Main Island and to welcome other geopark students to their islet home town as well. This exchange program has shaped the education of the involved schools in an exciting way. As the school teachers and students experience the peak experience of learning by sharing and by doing, they tend to become motivated to design locally appropriate teaching materials. It was found out that pedagogical materials, tailored for the local context, can easily provide students with a peak learning experience, because they reinforce complex feedback mechanisms. With such practice, students become more confident on guiding their cohort guests from other geoparks. They are proud to act as geopark ambassadors to visitors (Figs. 5.16, 5.17, 5.18 and 5.19).

Faced with contemporary unpredictable environmental change, the students are also being prepared to improve their observational and analytical skills to understand the changing environment. This can partly be taught in the classroom, but skills that are taught in the outdoor, natural environment itself are more likely to be effective. Therefore, it is believed that the exchange program for geopark students will produce more and more benefits as it proceeds. Because of that structural environment educational policy, support has to be continued and sustained by local communities, school teachers and government agencies (Ley et al. 2017). It is fair to say that geopark promotion and implementation in Taiwan has achieved a level where both top-down and bottom-up mechanisms meet with great success.

5.4 Challenges of Geopark Environmental Conservation

A variety of methods is used to conduct environmental education for conservation and educational purposes in Taiwan's geoparks. Pamphlets, posters, explanation boards, films and guided tours are the most typical methods used by interpreters and to encourage self-guided visits. There are increasing applications of mobile devices with APP software. The use of QR codes for visitors to scan for more site information has become popular as well. Because technological devices are becoming more common and visitors are better equipped for self-guided site interpretation and learning, promotion of environmental education through geoparks seems to be promising (Figs. 5.20 and 5.21).

However, with both the top-down structural support and bottom-up initiatives and mechanisms, the geopark promotion related to environmental education is faced with many challenges (and threats). They may not be instantly

Fig. 5.16 Students from various schools discussing issues at Matsu Geopark

Fig. 5.17 A student explaining a light house heritage site to visitors at Matsu Geopark

detrimental, but for long term and sound future geopark development, one has to act with caution.

(1) *Similarity within and among geoparks*

With tight networking among all Taiwan's geopark communities, close cooperation has been something that one should be proud of. Due to the diverse geo-environments of all geoparks, however it is considered creative and better to design distinctively different strategies of developing environmental education materials and methods. Close networking resulted often in relatively similar environmental education material, but creative, innovative collaboration is needed also to develop distinctive means of improvement.

(2) *Funding sources and community identity/autonomy*

Funding is a difficult issue. At the moment, most funding for environmental education in Taiwan's geoparks come from the national or local public sectors. Very little is from private

Fig. 5.18 Students holding a pamphlet explaining landforms to exchange visitors at Matsu Geopark

Fig. 5.19 Students of Taiwan main island geoparks playing saxophones to welcome Matsu students at Tsaolin Geopark

NGOs, individuals, or trusts. Although Taiwan is undergoing a national educational reform and many new policies are being developed, the lack of funding sources for geopark promotion can sometimes constrain ideas and methods of presentation. For a much freer and more creative development of geopark environmental education, more different funding sources, including those from the general public, may be vital, but also raise conflicts of interest.

(3) *Schools and local communities*

Schools play an important role in each geopark. They are where most adults would work for societal development and community future. If geopark activities are initiated at and for schools, the local communities usually work hard to make this a success. However, there are two extremes that may rise. One is when the geopark community interacts too

Fig. 5.20 Information boards at Matsu Geopark

Fig. 5.21 An explanation board at Matsu Geopark

little with local schools, and the other one is when the geopark community interferes too much in the school curriculum. It is essential to manage with wisdom the situation. When geopark communities strengthen their capacity to this end, a new level of balance will emerge. In case a geo-running event in a geopark with participation of schools is organized, it is significant how many of the children's parents will take part and thus demonstrating which balance has been achieved. Even better, if teachers thoroughly understand the values of geoparks, as a tool for education, it is possible that the school's home education could incorporate the environmental values of geopark. The geopark community will also understand more of the local schools education and seize the opportunity to engage in a productive way. In-service training for school teachers can also be structured by the national Ministry of Education.

References

Chen, Y. H. (2017). A geopark filled with life and death stories. In *Taiwan on geological scale: Tracing Taiwan geology from geoparks* (pp. 148–171). Taipei, Taiwan: Metropolis Publisher. (in Chinese).

Chiu, Y. Y. (2017). A Muddy Geopark with sweet fruit. In *Taiwan on geological scale: Tracing Taiwan geology from geoparks* (pp. 172–209). Taipei, Taiwan: Metropolis Publisher. (in Chinese).

Chuang, R. L., & Wang, F. (2017). A Sonata of Geopark with volcanic activites. In *Taiwan on geological scale: Tracing Taiwan geology from geoparks* (pp. 248–307). Taipei, Taiwan: Metropolis Publisher. (in Chinese).

Ley, S. Y, Chuang, R. L., & Wang, W. (2017). An epoch of geopark of iron and tears. In *Taiwan on geological scale: Tracing Taiwan geology from geoparks* (pp. 308–355). Taipei, Taiwan: Metropolis Publisher. (in Chinese).

Lin, S. H. (2013). The development and status Quo of Taiwan's environmental education. *Ecological Taiwan, 41,* 6–13. (in Chinese).

Shen, S. T., & Lin, S. F. (2017) A Cape of Geopark. In *Taiwan on geological scale: Tracing Taiwan geology from geoparks* (pp. 128–147). Taipei, Taiwan: Metropolis Publisher. (in Chinese).

Su, S. J., & Lin, J. C. (2014). The environmental and social implication of geoparks. *Geology, 33*(1), 56–69. (in Chinese).

Wu, L. C., & Chang, Z. C. (2017). The development of the accreditation of the public and private environmental education facility. *Journal of Environmental Education Research, 13*(1), 99–136. (in Chinese).

Wu, L. C., & Chang, Z. C. (2018). The hours, contents and methods of Taiwan's civil servants' environmental education learning activity. *Journal of Environmental Education Research, 14*(1), 1–37. (in Chinese).

Wu, L. C., & Wang, H. J. (2012). A positive study of accreditation environmental education facility. *Journal of Rural Leisure Studies, 6*(1), 1–13. (in Chinese).

Taiwan Geomorphology Study Group. (2013). The environmental education law and landscape conservation. *Communication for Landscape Conservation, 36,* 4–7. (in Chinese).

Yeh, H. C. (2017). Analyzing the context for the development of environmental education and sustainability education. *Journal of Environmental Education Research, 13*(2), 67–109. (in Chinese).

6 Geo-hazards and Geo-tourism as Stimulators for Geoparks

6.1 Geo-hazards as a Background for Geoparks

Taiwan's geology and geography demonstrate its critical environmental sensitivity and vulnerability. Out of the core values of geoparks, landscape conservation, environmental education, eco-tourism and local engagement, the previous chapter has introduced how environmental education is particularly vital for the society at large and for the local geopark communities. However, the understanding of geology and geomorphology is for some people unfortunately limited and often influenced through their empirical knowledge about various natural hazards. The collective experience of various geo-hazards in Taiwan has facilitated the formation of some legal structures for conservation and education for the environment. For geopark communities, the need of a deep and extensive understanding of their own environment is acute. For the obvious reason of combating and/or preventing potential natural hazards to become disasters, communities are alert and in acting mode. Therefore, it is fair to say that geo-hazards are negative stimulants which promote the proper understanding and use of a geopark's environment.

We will introduce four major types of natural forces that bring geo-hazards to the island state of Taiwan, including tectonic movement/uplift, earthquakes, typhoons and landslides. They play varying roles in all parts of Taiwan, sometimes intertwine with each other, thereby shaping the even complicated processes and are altering the landscapes and burden the society with natural disasters. These hazards may occur with the coupling effect of extreme and/or dramatic climate change which can magnify the damage to the infrastructure.

For example, based on long term typhoon research, the Lichi Geopark, located in the eastern coast of Taiwan, has a high probability to be affected by typhoons, even by the center of a typhoon. In 2016 solely, two typhoons crossed the county and the area of Lichi Geopark was seriously damaged. The extremely strong and high winds destroyed countless agricultural plants and crops in a short time and an extremely serious financial loss occurred. As these hazards have been constant, the society is faced with the need of seeking ways of adaptation and adjustment. How one copes with natural processes and disasters and how the community adjust or even transform its way of life, are big issues for the Lichi Geopark's farmers, depending on the land for their living.

6.1.1 Tectonic Movements

Recent studies have exploited the strong coupling effect between tectonic uplift/subsidence and erosion in process-based models involving drainage network evolution and hillslope processes (Allen and Hovius 1998; Burbank et al. 1996; Burbank and Piner 1999; Dadson et al. 2004; Hovius et al. 2000). Taiwan is located at the colliding boundary of the Philippine Sea Plate and the Eurasian Continental Plate. As a result, central mountainous Taiwan experiences an annual uplift rate between 2 and 4 mm. The rate in the north of Taiwan was 2 mm/year from 5500 year BP to 1500 year BP (Peng et al. 1977). The Philippine Sea Plate is converging towards Taiwan island at a rate of approximately 7–8 cm/year in a northwest direction (Seno et al. 1993; Yu et al. 1997). This tectonic setting has induced uplift rates of 5 ± 0.4 mm/year in the Central Mountain Ranges over the past 9000 years. It is due to such a tectonic background that the island of Taiwan is interiorly separated by several large structural lines and as a result most of the rock mass is relatively fragile.

Such dynamic environment resulted in many fault lines across the island and tectonic movements have a short occurrence period. The Chi-Chi earthquake might be the most disastrous event ever recorded and is a great example to educate the people on the significance of tectonic structures. The Chi-Chi earthquake occurred on the 21st of September 1999 with a magnitude of 7.3 on the Richter scale. During the Chi-Chi earthquake both horizontal and vertical movements occurred resulting in many fractures

J.-C. Lin and S.-J. Su, *Geoparks of Taiwan*, Geoheritage, Geoparks and Geotourism,
https://doi.org/10.1007/978-3-030-04895-2_6

Fig. 6.1 Landslide hazards after earthquakes and typhoons

along the Chelungpu fault and caused a relative uplifting between 1-9 meters along the fault line area (Fig. 6.1). There is not much detailed information on serious earthquakes of the past; however it has been proven that Chelungpu is the most active fault in Taiwan with a re-occurrence period of less than 150 years.

Tectonic activities, movement and fault lines, contribute to the making of many landscapes and the parks have been chosen to represent this. For example, at the Northern Coast Yehliu Geopark, there are many fault-related landscapes for visitors to explore. There is even evidence of tectonic uplifting where a vertical sea notch is revealed (Fig. 6.2). Sea notches and terraces formed by tectonic uplifting then become excellent evidence for geo-education on dynamic tectonics.

Tectonic uplifting has produced a series of sea notches on the eastern coast of Taiwan. Volcanic activities are also a result of tectonic activities. This is particular true in the northern part of Taiwan where a lot of fumaroles still produce sulfur smoke and sizzling hot sulfur springs. Hot

Fig. 6.2 Vertical sea notch as result of tectonic uplifting

Fig. 6.3 Post volcanic activity in northern Taipei showing fumaroles with sulfur smoke

springs are also valuable resources for recreation and health therapy (Fig. 6.3). All of these can create diverse landscapes and are of economic value despite bringing many geo-related hazards to the geoparks.

6.1.2 Earthquakes

Since the Chi-Chi earthquake of 1999, most of central Taiwan has become more fragile than before. Therefore, it is fair to say that the property damage, as well as agricultural and other economic losses became all immediate cost (Fig. 6.4). More detrimental and costly damages are the ones that are linked with other geo-hazards, like landslides and rock falls that are common among environmental hazards. A few large earthquakes, coupled by several typhoons, torrential rains and even regular rain, are contributing to a changing environment with complicated hazardous threats.

In a tectonically active area like Taiwan, landslides triggered by earthquakes or heavy rainfall events are common

Fig. 6.4 A slope failure after a strong earthquake

Fig. 6.5 A scene of a damaged road after typhoon

Fig. 6.6 Trails in mountain areas are easily damaged by heavy rainfalls

features. During the Chi-Chi earthquake extensive damage could be attributed to landslides in the central part of Taiwan. In Nantou county more than 21,000 landslides occurred, and in Taichung 650 landslides where registered along the Central Cross Island Highway in central part of Taiwan (Figs. 6.5 and 6.6). These geo-hazards, when coupled by other natural forces or incidents, could lead to compound hazards throughout time.

6.1.3 Typhoons

The precipitation brought by typhoons is one of the major sources of rainfall for Taiwan (Hung 2009). However, these rainfalls are often so strong and intense that the hazards triggered are of great magnitude and cause spectacular damage to the society and infrastructure. During a typhoon, the windward slopes may receive more than 1000 mm of

Fig. 6.7 Soil erosion at Lichi Formation's mélange mudstone at Lichi Geopark. Careful land use is required here

rainfall within a 24 h period. That certainly becomes exceptionally detrimental where fault lines are regularly impacted. The local community of a geopark could volunteer with in situ guardians and monitors for environmental changes, and, thus, revealing a major bottom-up power.

Along this line it can also be demonstrated how human activities can exacerbate the natural processes. Landslides can be caused by a poor combination of pathways, trails and settlements. Experience also demonstrates which safety measures are successful and which are not economically worthwhile (Fig. 6.6).

As pointed out earlier, the Lichi Geopark is a geopark where typical typhoons would strike due to its location and the fragmentation of landform becomes of concern. Not only its sedimentary mélange landforms are changed drastically (Fig. 6.7) but also its local agriculture and economy are detrimentally influenced during typhoon seasons. Therefore, in order to cope with the natural processes of typhoons, besides people interested in landforms, also the farmers whose livelihood depend on a stable farming environment have to be involved. A mission for the geoparks and administrations is to ensure that the local people's understanding of geo-hazards and their physical and climatic mechanisms contributes to their wellbeing.

6.1.4 Landslide and Mass Movements

There are landslides and mass movements in some geoparks of Taiwan, especially in the mountainous areas. Mass movements together with earthquakes and typhoons make Taiwan an island which suffers a high rate of erosion.

Mass movements are often interchangeable with the term slope collapse. The slopes undergo processes such as weathering, disintegration and failure. All slopes are influenced by constant gravity. As a result, the weathered geological materials, such as soils and rocks, slide downward along the slope. Types of landslides include debris flows, rock sliding, soil sliding, mudslides and rock falls. Many areas need to increase their environmental resilience to face these natural forces and human influences by knowing the environment and making sustainable decisions. Figure 6.8 shows a dramatic scene of rock falls out of basaltic columns at Penghu Marine Geopark. This picture points to the importance of safety for visitors and the significance of site management as the geopark becomes a popular destination for both national and international tourists.

Debris flow is another phenomenon that often comes along with earthquakes and rainfalls. The necessary requirements of a debris flow include abundant amount of geological materials, special landforms, sufficient amount of rain, an appropriate angle of (valley) slope, and suitable geological structures. These are very common conditions in the mountains of Taiwan. The power of landslides is often compounded by the presence of human structures that range from house damage to loss of life (Fig. 6.9).

The majority of soil erosion, debris flow and landslide processes in Taiwan are the result of a combination of natural and human induced processes (e.g., Lee 2014; Chen and Chang 2002). However, combined with the increasing use of marginal lands and steep slopes, soil erosion and soil sliding are accelerated by human activities. Slope failures are triggered easily (Fig. 6.10). This has in turn resulted in decreased reservoir capacity, river channel damage to

Fig. 6.8 Rock falls of basaltic columns at Penghu Marine Geopark

Fig. 6.9 Landslides caused damage to a house at Lichi Geopark

communication infrastructure and impacts on the local flora and fauna. A holistic approach to land use planning, therefore, is necessary. This must involve a national classification of the land according to "fit-for-purpose" criterion (Lin and Su 2013). In geoparks, these issues, related to global environmental change, have thus become focused and policy implications are now based upon global perspectives.

For example, when the Tsaoling geopark communities understood the potential threat of detrimental and hazardous landslides, local solidarity in facing and understanding the natural forces became a communal learning object (Chen 2017; Lee 2011, 2013). That proved that the community adopted a learning attitude and became humble in front of the powers of nature. The local community of Yenchao geopark knew that soil erosion of the bad land area was bad for the local agricultural economy, but with an increased understanding of the processes sculpturing their land, the local people gain optimism concerning the future of their own unique land and its capacity to produce useful crops. Further, a new aesthetic capacity was generated that appreciated the unique beauty of the bad land landscape and resulted in a strong local community response (Fig. 6.11).

Fig. 6.10 A scene of slope failure in the conglomerate badland of the Sanyi Hoyenshan Natural Reserve

6.2 Geo-tourism for Local Sustainability

The steady growth of tourism and related economy has encouraged unique or niche products in this world of diversity. Nature-orientated tourism is experiencing an annual increase of 10–30%. It is estimated that there will be 1.8 billion international tourists by 2030 (United National World Tourism Organization 2018). The rapid growth of tourism and tourists brings economic benefits but also causes environmental impacts that might not have been recorded or experienced before. The World Commission on Environment and Development (WCED) and the International Union for Conservation of Nature (IUCN) support the concepts of sustainable tourism and eco-tourism. So that people,

Fig. 6.11 Rill and gully mudstone erosion in Yenchao Geopark

natural resources and recreation can achieve a positive balance for sustainable development.

In July 1998 the United Nations Economic and Social Commission held its forty-sixth General Assembly and designated the year of 2002 as the International Year of Ecotourism. The World Tourism Organization and United Nations Environment Program jointly promoted ecotourism as a strategy for development, with the goal of maintaining biodiversity and the hope of promoting global participation, review, and exchange of good practices. Through these exchanges, it was an attempt to construct an ecotourism model that allows tourism to promote economic prosperity while balancing environmental protection and sustainable development.

The endeavors to organize geoparks in Taiwan are considered to be a practice that is in line with the ecotourism and sustainable development concepts promoted by the United Nations Economic and Social Commission, the World Tourism Organization and the United Nations Environment Program. Geo-tourism is indeed crucial in geopark development as many geoparks are located in places with less advantageous conditions for economic activity and perhaps with more challenging physical conditions.

As mentioned in previous chapters, the *Environmental Education Act* of Taiwan has been a policy facilitator *for* the environment and *of* the environment. While some environmental education programs take place in indoor classrooms, it is realistic for environmental education to take place *in* the environment that facilitates the education and deepen environmental experience for education to become better personal experience. It is worthy of notice that it was the geoparks themselves that started the practice of local geo-tourism. In today's fashion for slow living and pursuit of unique local experiences, geo-tourism can encourage sustainable local development in which local people's own stories and observations of local environment and life are valued. This in part correlates with the geopark value of local participation and further ensures the sustainability of geo-tourism. Each geopark of Taiwan has its niche for different types of tourism, such as hiking, diving, biking, observing, tasting and even gardening. Promoting geo-tourism will also reveal the potential of each local landscape for further understanding of these aspects.

By attracting visitors to geo-sites, geo-clusters or/and geoparks of particular beauty, we are not only developing geo-tourism activities based on geology and landscapes (Chiang and Lee 2014), but also fulfilling the local economic needs and empowering the local community with local environmental knowledge and interpretation skills. Such knowledge-oriented capacity building not only enhances the local economy, but also improves our guardianship of the local resources.

Examples from the Southwestern Coastal Wetland Geopark (i.e., Yunchianan) provide an illustration of how such web of protection could be formed. The little village of Shi-tsao in the An-nan district of Tainan city is an historical site where the Dutch colony and its battles with the Qing dynasty for the coastal territory and resources took place. But as the historical accounts have been conflicting about the historical battles, there is one clear thing that has attracted one to two million visitors per year from around the country. It is the combination of mangrove ecology and green canopy water that becomes attractive. Education about mangroves takes place during a romantic boat ride under the shady green canopy in a short canal, believed to be built by the Dutch for defense reasons (Fig. 6.12). During this journey visitors learn the five species of mangroves along the water way. The green canopy is now becoming well known as the peak experience of learning. Another scene at the same geopark is the occurrence of salt fields. Salt production was manually done at this southwestern coastal area but due to economic reasons and shortage of labor, salt production has become a thing of the past. The salt fields have been turned into tourism and leisure areas. They are now more important than before and are a critical component of this coastal facility. Local interpreters, environmental education and local story telling are all gathered around these historical fields (Fig. 6.13).

Another issue of concern for geo-tourism is carrying capacity. While the concept of carrying capacity has a relative standard definition about its concern of not exceeding a destination's limit on number of visitors and no adverse consequence should be brought, it is a contentious concept. Some cautions and critiques about its lack of dynamic consideration are taken serious (e.g., Buckley 2000; Wagar 1974). In particular, if we take the physical carrying capacity of a destination to compare to its economic carrying capacity, social carrying capacity, and even biophysical carrying capacity, it will be clear that the concept needs contextually operational definition and organically realistic representation.

That being said, "carrying capacity" for many Taiwan geoparks are not easily exceeded because they are not usual tourist sites, most of them are not popular tourist sites. However one typical example of carrying capacity is that of the Yehliu Geopark. The geopark is famous for its coastal erosional landscapes. With intriguing landforms (Figs. 6.14 and 6.15), Yehliu Geopark has attracted almost 3,000,000 visitors annually in the last three years. As the increase of visitors has been dramatic, from half million in 2005 to 3 million, today the carrying capacity of the geopark site needs realistic assessment. Policy making towards maintaining a good carrying capacity must be developed for better site management. It is even part of the policy for the mitigation

Fig. 6.12 Green canopy covered waterway surrounded by mangroves at Yunchianan Geopark attracts many visitors

Fig. 6.13 Old salt fields turned into tourist attraction site at Yunchianan Geopark

of geo-hazards. When the site is crowded with more than the bearable population, the erosive coastal sandstones are threatened and in turn become a threat to the visitors. When visitors do not take caution to protect themselves from slipping into the ocean, tragedy may happen. For better management practice of the geopark, when the visitor number is over 2500 people per hour, the geopark entrance is temporarily closed. Visitors are directed into the Visitor Center of the geopark to view multi-media films or to interact with their interpreters while waiting for their turn to enter the park.

Geo-tourism has an integrative effect on the operation of the local economy. The Penghu Marine Geopark showcases such forward and backward economic linkages. The Penghu archipelago is located off shore to the west of Taiwan Island, in the strait between Taiwan and China. As Penghu is

Fig. 6.14 Mushroom rocks are major tourist attractions at Yehliu Geopark

Fig. 6.15 cd rocks are characteristic in Yehliu Geopark

composed of so many islands, its geo-tourism and leisure activities among the islets have to be connected and served by boats (Fig. 6.16). The boat trips and scheduling become a way of controlling how the geo-tourism of various islets is carried out. For islets with a frequent boat trip schedule, visitors tend to stay briefly, but for islets with less frequency of boat trips or even without regular public boat transportation visitor numbers tend to be lower but they stay longer. Therefore, all islets show in-depth and responsible geo- and eco-tourism.

The rich geo-diversity and bio-diversity of the archipelago is very attractive. But the cultural background of such environments are as well and even more attractive as they are full of 'stories'. The people often take pleasure in showing visitors about their life in the windy and infertile soil environments. For instance, cactus is a plant with little need of water and not much demand from the soil in which they grow. Cactus fruit have become important ingredients for making ice cream and icy popsicles for tourists. The locals have invented a special device to pick prickly pear

Fig. 6.16 Geo-tourism by cruise and boats is now gaining popularity in Penghu

fruits so that the pickers would not be hurt by the thorny fruits on the bush. Such invention of tools is certainly based on the local experience of picking the thorny cactus fruit. This experience adds a unique interest to the tourists' visit and ignites a memorable insight into the life on the island (Fig. 6.17). Good interpretations serve the right purpose of bringing the locals' lived experience to the visitors (Fig. 6.18).

Basaltic columns in different shapes are common on all the various islets. The local people used to utilize the basaltic column to build houses and temple structures, constructing fishing weirs, and erecting short walls for vegetable gardens. Many human settlements and human activities are supported by the utilization of basaltic rocks and columns. Different basalts are characteristic of the islets and they are part of the archipelago culture. Basalt sculptures are some of the most beautiful of the Penghu products and could become very famous and treasured (Fig. 6.19).

Clean ocean water and ocean atmosphere is essential to the archipelago not only for geopark promotion, but also for

Fig. 6.17 A woman picking cactus fruit with a special tool at Penghu Marine Geopark

Fig. 6.18 Geo-tourism with good environmental interpretation is an integral part of Penghu Marine Geopark

the livelihood of the people of the archipelago. The fishing economy is still important for the locals; the visitors who come are often looking for fresh and good catches and clean water to dive in (Fig. 6.20). It is thus critical that the marine geopark is maintained by the local initiative. Protecting and conserving the marine environment of the Penghu archipelago is a first priority for geopark development with good local initiatives and supports can be a substantial tool for sustainable development of the marine communities.

Geo-hazards and geo-tourism are two important aspects of geoparks. They are, however, both stimulators for geoparks, one as negative and the other one as positive stimulator. In a philosophical perspective, the two are depending on each other, like yin and yang. They are the two sides of a geopark. Geo-hazards are the very condition that prompts people to take control of the environment and the community to manage their own environment. Geo-tourism is then a mechanism to make the most appropriate and sustainable use

Fig. 6.19 Basaltic columns are used for the logo of Tongpan geosite in the Penghu Marine Geopark

Fig. 6.20 Overlooking the tranquil sea from the Hujin geosite of the Penghu Marine Geopark is a peak experience for visitors

of the environment. In geoparks, matters are interdependent on one another. It is a web of network in geoparks. A harmonic interdependent network of human and non-human agencies is where geoparks base upon.

References

Allen, P. A., & Hovius, N. (1998). Sediment supply from landslide-dominated catchments: Implications for basin-margin fans. *Basin Research, 10,* 19–35.

Buckley, R. (2000). An ecological perspective on carrying capacity. *Annals of Tourism Research, 26*(3), 207–210.

Burbank, D. W., Leland, J., Fielding, E., Anderson, R. S., Brozovic, N., Reid, M. R., et al. (1996). Bedrock incision, rock uplift and threshold hillslopes in the northwestern Himalayas. *Nature, 379,* 505–510.

Burbank, D. W., & Piner, N. (1999). Landscape evolution: the interactions of tectonics and surface processes. *Basin Research, 11,* 1–6.

Chen, Y. H. (2017) A Geopark filled with Life and Death Stories, in *Taiwan on Geological Scale: Tracing Taiwan Geology from Geoparks*, pp. 148–171. Taipei, Taiwan: Metropolis Publisher. (in Chinese).

Chen, Y. F., & Chang, F. N. (2002). *The furious landslides of Taiwan in the 21st century: Natural disasters and human disasters.* Kaohsiung, Taiwan: Pioneer Press.

Chiang, W. C., & Lee, J. F. (2014). Geo-tourism in Nei-hsuang River. *Geology, 33*(1), 66–73. (in Chinese).

Dadson, S. J., Hovius, N., Chen, H. W., Dade, B., Hieh, M. L., Willett, S. D., Hu, J. H., Horng, M. J., Chen, M. J., Stark, C. P., Lague, D., & Lin, J. C. (2004). Links between erosion, runoff variability and seismicity in the Taiwan Orogen. *Nature, 426,* 648–651.

Hovius, N., Stark, C., Chu, H. T., & Lin, J. C. (2000). Supply and removal of sediment in a landslide-dominated mountain belt: Central range, Taiwan. *Journal of Geology, 108*(1), 73–89.

Hung, C. W. (2009). The precipitation patterns of Taiwan: A perspective of 100 years of historical meteorological data. *Taiwan Historica, 60*(4), 45–69. (in Chinese).

Lee, C. T. (2011). *The development process of Tsaoling Geopark.* Taipei: Forestry Bureau of Taiwan and National Taiwan University co-published. (in Chinese).

Lee, C. T. (2013). *The transformation of Tsaoling's geomorphology.* Taipei: Forestry Bureau of Taiwan and National Taiwan University co-published. (in Chinese).

Lee, C. T. (2014). Landslides and Tsaoling Geopark. *Geology, 33*(1), 31–35. (in Chinese).

Lin, J. C., & Su, S. J. (2013). *Landscape conservation of Taiwan.* Taipei, Taiwan: National Taiwan University. (in Chinese).

Peng, T. H., Li, Y. H., & Wu, F. T. (1977). Tectonic uplift rates of the Taiwan island since the early Holocene. *Memoir of the Geological Society of China, 2,* 57–69.

Seno, T., Stein, S., & Gripp, A. (1993). Philippine sea plate consistent with NUVEL-1 and geological data. *Journal of Geophysical Research, 98*(bi0), 17941–17948.

United National World Tourism Organization. (2018). *International tourists to hit 1.8 billion by 2030.* http://media.unwto.org/en/press-release/2011-10-11/international-tourists-hit-18-billion-2030. Accessed at August 10, 2018.

Wagar, J. A. (1974, May). Recreational carrying capacity reconsidered. *Journal of Forestry*, 274–278.

Yu, S. B., et al. (1997). Velocity field of GPS stations in the Taiwan Area. *Tectonophysics, 274*(1–3), 41–59.

7 Landscape Conservation as a Tool for Sustainability

7.1 Landscape Conservation

Landscape conservation can be conducted with different methods and approaches that depend on the concepts and perspectives adopted (Lee et al. 2010; Wang and Ho 1999; Wang and Lee 2002). For instance, the level of technological and societal understanding determines how landscape conservation is implemented. Most commonly, landscape conservation is understood as a way to protect the landscape from inappropriate or unfit changes. However, it also concentrates on maintaining sustainable ways of human life. Landscape conservation does not mean that it tries to keep the landscape unchanged. Ideally, the value of landscape conservation also resonates with global environmental changes and human societal challenges.

Landscape conservation not only depends on the expectations and capacity of society, but is also determined by the holistic values which yield a better quality of life. Furthermore, given that each environment is different, each society has its contextually appropriate ways of conducting conservation. In Taiwan, the official and structural umbrella for landscape conservation was formally started in 1984 when the *Cultural Heritage Act* was passed by the government. In the *Act*, natural reserves and natural sites with scientific value were advocated. Since 1984, various nature reserves have been designated. Later in 2016 when the geoparks, natural heritages sites, and natural monumental sites are formally incorporated into the amendment of the act, the act becomes inclusive of both cultural and natural objects (Wang 2017).

The Houyenshan Nature Reserve, for example, is one of the first of the eight nature reserves developed in 1986 under the previous Cultural Heritage Act (Fig. 7.1) (Forestry Bureau of Taiwan 2018). Among the first eight Nature Reserves of Taiwan, the protection of flora and fauna was most important. Houyenshan Nature Reserve was the only and first reserve in which its landslide and cliff landscapes were highlighted and deemed as important to be studied and conserved for further investigation and education purpose. After continuous scientific studies of the Houyenshan conglomerate area for more than three decades, it has been proven that the area shows an exciting environmental history and displays beautiful riverine environmental changes. Recently, the Taiwan's Forestry Bureau has conducted a series of scientific studies on Houyenshan as well as a program for local education on the particular issues and problems with conglomerate rocks and bad land geomorphology. This is a case of how landscape conservation has been transformed by the locals to some observation initiatives that have shown that effective, scientific research, environmental education for the public, protection tools, and community monitoring are keys for success. Today, Taiwan has 23 Nature Reserves under the national *Cultural Heritage Act*.

In July 2016 the national *Cultural Heritage Act* successfully added an amendment, which is critical for landscape conservation. The amendment covers inclusive elements concerning natural and cultural landscape conservation. One concerns physical and geological sites of significant scientific and natural value; the other includes historical and cultural heritages. Both are particularly important, de facto, for collaborative activities of all stake holders devoted to conservation. It certainly encompasses the integration and advancement for trans-disciplinary and trans-sectoral cooperation of both, private and public sectors as well as the utilization of bottom-up and top-down forces.

Proper contextual landscape conservation will ensure good opportunities for the appreciation of nature, outdoor recreation, environmental education, tourism and similar activities. It includes suitable maintenance and improvement of what has been inherited from nature, the history and culture of a region. Nevertheless the provision of new amenities and facilities has to be undertaken with care. The dynamic landscapes of Taiwan are rare and vulnerable, and some features are impossible to recover once destroyed (Lee 2016; Lin 1999, 2014; Lin and Su 2013a, b; Wang 1997). All valuable landscapes require care. Such landscape conservation can be achieved through education at all levels to generate a knowledge-based picture of what we have and

J.-C. Lin and S.-J. Su, *Geoparks of Taiwan*, Geoheritage, Geoparks and Geotourism,
https://doi.org/10.1007/978-3-030-04895-2_7

Fig. 7.1 Erosional catchment in the conglomerate of Houyenshan Nature Reserve

what we may lose. Understanding the interrelationship between geology, geography, environment and society provides a context for decision making. Techniques for sustainable management and maintenance of landscapes can then be chosen, and even invented for each unique situation and settings.

Landscape conservation schemes require strong partnership between government and citizens. The schemes need to address issues at the community as well as at the national level in order to be sustainable. One part of the geopark mission in Taiwan is to conserve the environment and to revitalize local socio-economy so the role of geoparks is crucial for the hosts as well as the nation state.

Nowhere in the world one can find an island of this size with such rich diversity. The climate, geology, geomorphology and geography interact closely to form versatile habitats that become home for many species and various cultures. Diversified terrains, seasonal typhoons and sub-tropical climate define this milieu. As a small marine country and with scarce resources, the proper use of resources and management of the nature is crucial for a sustainable Taiwan. Only a quarter of Taiwan's land is good for agriculture and habitation. The huge population of 23 million places great pressure on its land. Urbanization, overuse of slopes and marginal lands, such as tidal land and riverbeds, are just a few examples to illustrate the stressed island ecology. When these limits are pushed, the ecological environment and the cultural heritage are endangered.

7.1.1 The Conservation of Cultural Landscapes

Cultural landscapes are now created using complex techniques and technologies. Both humans and habitats then adapt and adjust to each other and the result is the complicated environment that we live in.

An example such interactions can be seen in the network of underground tunnels in Matsu. The tunnels were excavated by the military personnel during the cold war period (see Sect. 1.2.6 and 5.3.3) and have now become a military-driven landscape and a cultural heritage.

Another example of cultural landscapes that have an important economic history are the aquafarms for abalones in the Northeast Coastal Geopark (Fig. 7.2). The abalone aquafarms were constructed by local fishermen in the coastal areas close to their homes. Those constructions were created at a time when abalone had a high economic value and when the coast was not regulated for nature conservation. Although most of the aquafarms are not allowed to operate today as the coastal laws and regulations are implemented with more scrutiny and care. Artificial structures and buildings set up by aquafarmers are considered illegal. The once profit-making aquafarm structures become a part of the economic cultural legacy.

The geo-diversity of Taiwan includes both the nature and the human culture. Understanding and respect for the interaction within the two extremes is a solid foundation for sustainability. Hills and plains, mountains and coasts, and

Fig. 7.2 Abalone aquafarms produce high economic value, but they destroy natural coastal landscape

basins and tablelands all bear the hallmarks of human activity (Lin 2015). Cities and towns are created amidst them. The natural environment has been modified to meet the food, shelter, societal, cultural and communication needs of humankind, but inappropriate land use and developments in environmentally sensitive areas often trigger hazards that are beyond control. Steep slopes, tidal flats, riverbeds, and areas subjected to subsidence are typical unsuitable sites for human activity or development. Such developments not only bring harmful results to the environment and ecology, but also to the human settlements. This costs loss of life and damage to properties. It is now an urgent matter for Taiwan to develop new programs for sustainable management.

The surveys and audits of the environment should identify useful measures for the maintenance of environmental quality (Wang 2017). Taiwan has achieved strong economic growth but often at the cost of the environment. It is now appropriate for the central and local governments, government agencies, schools, businesses, communities and individuals to subscribe to the landscape conservation agenda through specific geopark-related activities.

7.2 Landscape Conservation and Sustainability

Landscape conservation is critical for the sustainability of the landscapes of Taiwan. Natural hazards are often the result of the compound effect of poor environmental management and social practices (see Chap. 6).

7.2.1 Environmental Change

Global environmental change is of critical significance to Taiwan because of its island, sub-tropical location, geological condition and geomorphological processes. Typhoons often induce high intensity rainfalls and floods. Too much water is often followed by droughts which pose particular challenges to the management of soil and water resources and to health and food safety. These issues warrant the need for a coordinated national resources allocation and development plan. In facing and managing the rapidly changing environments, both international and national environmental policies play big roles, Local endeavors are also crucial to the formation of bottom-up initiatives to safeguard the environment. Here the geopark organizations can be a useful tool.

In the face of extreme weather events, river and slopes are areas where problems regularly occur. Rivers in Taiwan are generally short and with steep channel gradients. They are capable of transporting large volumes of sediments (Fig. 7.3). With extreme rainfalls, rivers could burst their banks and overflow surrounding areas, burying productive agricultural land and buildings in its path (Fig. 7.4). River management and engineering is often called for to reduce or alleviate flooding. Thus, it is important for the local population to be aware of climate change and extreme rainfalls, and to try to reduce the damages caused by hazards. Understanding the vulnerability of our environments and the risk of our society, settlements, households and even each individuals is critical. With such awareness or caution of risk, better decision making and political monitoring can become a focus for the society.

The majority of soil erosion and landslide events in Taiwan are usually the immediate results of heavy rainfalls. However, with increasing pressure to utilize marginal lands and slope lands, human activity and developments have accelerated erosion rates and triggered slope failures. These have resulted in decreased reservoir and river channel capacity, damage to major and basic infrastructure (Figs. 7.5 and 7.6) and impacts to the local flora and fauna. The concept of landscape conservation is part of the crucial basics for a holistic approach to land use planning. The classifications of the land according to "fit-for-purpose" criterion should come out of these goals of sustainable development. All of these failures could have been avoided by a deeper understanding of the environmental processes and the use of applied geomorphological mapping and conservation measures.

Another extreme of environmental change is water scarcity that has an important political, economic and social impact. Taiwan has an average annual precipitation of 2200–2500 mm, which is mainly brought by monsoons and typhoons. Water conservation schemes integrated with landscape conservation are necessary in order to ensure that supply meets demand. However, monsoons and typhoons are seasonal and thus extreme rainfall is a typical trigger for erosion, landslides and debris flows. All environmental hazards intertwine that makes it necessary to install a comprehensive conservation scheme. Again, systematic promotion of geopark education could be part of the solution.

7.2.2 Coastal Retreat and Conservation

Taiwan is an island state with a total coastal length of 1985 km (status of 2016). Along the coast line, only 55% of it is natural (Construction and Planning Agency, Ministry of the Interior of Taiwan 2017). The rest of the 45% is mainly artificial, meaning the area is reinforced with artificial construction works (Fig. 7.7). The high proportion of artificial coastal line emphasize issues of coastal loss and retreat. Coastal retreat is the result of erosion and weathering as coastal material is removed away from the coastline or there is a shortage of materials being washed from the upstream of rivers to the downstream coastal area. The coastal retreat or loss occurs when riverine material supply and the wash-away of material from a section of beach is not in balance. The deficit may occur instantaneously during extreme weather or marine surface events associated with storm surges. Long-term erosion reflects the deficits in the littoral drift budget, which is often caused by a deficit in the supply of material to the beach section in question. Erosion along the coastal zone causes damage to coastal roads, creates unserviceable ports and harbors, impacts on marine habitats, damages real estate and de-values recreational areas. Therefore, coastal conservation requires a systematic knowledge and the public understanding of the system in order to make any coastal conservation policy schemes sustainable. The geopark programme can be used to facilitate this.

Fig. 7.3 Huge volume of sediments deposit in an alluvial fan

Fig. 7.4 Transported sediments in fluvial channel caused serious hazards after torrential rains

Fig. 7.5 Debris flow could bury and damage buildings after typhoons

7.2.3 Earthquakes

Seismic events, whether natural or generated by humans, are commonly described as earthquakes, but the causes include the rupture of geological faults, volcanic activity, landslides, mine blasts, and nuclear tests, just to name a few. Taiwan has a high density of geological faults and frequently suffers from earthquakes (see also Sect. 6.1.2). The effects of earthquakes include incidences of landslides, increased weathering and erosion resulting from loosening of rock and slope soil, displacement of communities, disruption of road, rail and waterway networks, power and water supply outages, and damage to commercial and domestic properties. Inventories and surveys conducted before and after such hazards can be significant to make the effects of earthquakes known to the public. Policy formulation for the better management of potentially quake-prone areas can help to prevent and reduce the further catastrophic effects of earthquakes.

Fig. 7.6 Drifted woods after typhoons often buried human structures but they might be resources for art work in the hands of artisits

7.3 Geoparks and Landscape Conservation

7.3.1 Landscape Inventory as a Foundation for Geoparks

With its isolated marine environment in a sub-tropical zone, Taiwan has a huge biodiversity but limited natural resources. Conserving natural resources and landscapes thus becomes crucial in order to maintain environmental and social well-being. Landscape conservation is meant to protect our living environment, and to improve its quality. Landscape conservation is important for protecting both natural beauty and habitats, and soil and water resources. A quality living environment can only be achieved if the fundamental resources of land, water and air are properly cared for. To achieve this, it is essential to catalogue the resource base.

Over the decades, valuable Taiwan landscapes and limited resources were lost to rapid industrial and economic growth. The quality of the living environment was compromised. Improper environmental management for economic growth of the past has been re-examined and a new insight of environmental governance is being formed.

To safeguard the nature and improve the quality of the living environment, the Forestry Bureau of the Council of Agriculture launched a series of landscape inventorying projects starting in 1994 to survey, record, and register landscapes of particularly scientific, environmental and educational significance.

Fig. 7.7 Tetra-pods are often used to protect retreating coasts although they may be brutal solutions to the nature

The inventory project turned out to be the important foundation for the conceptualization of Taiwan's geoparks. Well managed long-term programs for conservational education were seen to be crucial steps. The local communities of geoparks have been convinced to adopt the geoparks values and tailor them to suit local situations. To use the Yenchao Geopark as one example, given its high erosion problems of the mudstone, the surveys have revealed that, because of the magnesium content of the mudstone, the land is capable of producing high quality guavas fruits. In this example the inventory project is seen to produce multiple and unsuspected benefits.

7.3.2 Conservation as a Tool Toward Sustainability

Conservation is not only about the caring of landscape for public enjoyment and opportunity for outdoor recreation, tourism and other activities. It also involves maintaining and improving what has been inherited from the Earth and from the work of generations of people. Now the present society works toward the provision of additional amenities and better living quality for future generations. As we have seen the dynamic landscapes of Taiwan are vulnerable, and most are impossible to recover if destroyed, they require care and attention. We have also shown that conservation can be achieved through education at all levels to generate the knowledge of what we have and what we may lose if actions are not planned and maintained properly. That is, the understanding of the inter-relationships between geology, geography, environment and society is essential, as is technical training.

For instance, the pseudo model of Princess Ponytail at Yehliu Geopark is one example of good management practice (Fig. 7.8). The sandstone mushroom rock Princess Ponytail at Yehliu Geoaprk has been visited by many visitors, as has the sandstone mushroom rock called the Queen's Head. The path to the Princess Ponytail however is narrow and crowded most of the time. It causes danger for visitors while passing the path along a slippery edge to the coast. In response to this situation the park has utilised advanced 3D scanning techniques to replicate the natural Princess Ponytail. This has been placed in a quieter garden space so that visitors can more easily enjoy the monument.

Landscape conservation schemes require strong partnership between governments and citizens. These schemes need to address issues at the community as well as at the national and global levels in order to provide a perspective that is robust and sustainable. Landscape conservation in Taiwan has been under the auspices of the Council of Agriculture for more than a quarter of a century. Professor Shin Wang, a geologist and geographer, was the first scholar to undertake the task associated with landscape conservation through inventorying, surveying, recording, and planning to support policies for national protected areas. The introduction of the concept of "landscape conservation" into the curriculum of high school geography was another big step in 2006. Along with inventorying landscapes, educating the youth to conserve landscape is thus well supported.

Fig. 7.8 The model of Princess Ponytail is a replica of a natural sandstone mushroom rock at Yehliu Geopark

For the past quarter of a century, Taiwan has undertaken landscape conservation in different ways. To achieve the aim for sustainable socio-economy and environment, the means and objectives have been constantly redefined. They are mainly as follows:

(1) To survey and study sites in order to establish a comprehensive inventory of landscape;
(2) To identify the problems and threats that exist at particular geo-sites;
(3) To evaluate, preserve and manage selected geo-sites;
(4) To raise awareness of landscape conservation by educating local population, educators and public servants;
(5) To build networks between the geo-sites and geoparks of Taiwan and promote international exchange;
(6) To establish a website to facilitate sharing research results and information for landscape conservation; and
(7) To adopt the core values of geoparks and share Taiwan's experience with and learn from international community.

In 2010, the landscape conservation of Taiwan reached a new horizon. The academic geopark team expanded its perspective to include socio-cultural and community elements by inviting social scientists to join the team (Su and Wang 2013). This brought the adoption of a bottom-up approach and sought the participation of local communities. This approach has been effective in terms of environmental education for the local communities and in terms of engaging communities to work with each other for better landscape conservation and interpretation programs. Such engaging of communities follows the practice implemented by the Global Geoparks Network (GGN), and it is entitled the Taiwan Geoparks Network (TGN). Further, the teaching and sharing of interpretation programs among geoparks has stimulated the demand for environmental knowledge. These environmental education programs have been provided by scholars with the support of the regional governments and various central agencies. All these have formed a web of good practice among geopark communities.

7.3.3 Capacity Building for Conservation

In the national landscape conservation project, it is important to build strong government departments or agencies. In-service training in landscape conservation is provided by an academic team of scholars, partly supported by the Forestry Bureau. Local communities are engaged to present their work during the network meetings and conferences. Such meetings, conferences and workshops become a platform for communication between government agencies, the geopark communities and the academics. It is best described as capacity building. Through the mechanisms of co-learning, more government agencies and local communities have now joined the academically initiated geopark teams.

Through it, the programs emphasize that the environmental living quality of the community can be planned and targeted for improvement. Legislation and codes of practice are then emphasized as the local communities become more aware of the environment. The uniqueness, scarcity and vulnerability of special landscapes of each geopark is thus understood and cherished. The hazard prone areas are identified and risk assessments for new developments are applied. Although Taiwan has a relatively strong system for environmental impact assessment (EIA), the local community's natural awareness is also very good. This process helps to establish a holistic environmental perspective.

According to the July 2016 amendment of the *Cultural Heritage Act*, unique and significant ecological sites and geoparks should be designated as protection sites. Any alteration or damage to the original status of the sites should be prohibited. Although the Forestry Bureau is in charge of the protection, maintenance and management of the geological sites and geoparks, the geopark communities are the ones who most benefit from the amendment because through working together with the law they become empowered.

7.4 Conclusion

"Ilha Formosa!" (translated as "a beautiful island") was what the Portuguese sailors exclaimed when they saw the island of todays' Taiwan, when sailing along the east coast in the 16th century. It is a slogan that those who understand the island's geo-, bio- and cultural diversity will appreciate. To sustain the diversity, contextual conservation and strategic management of the landscape and environment is seen to be a fundamental idea in Taiwan. Sustainable development becomes more than a slogan. It is about care, action and persistence to look after the environment. Today human activity is increasing and we are placing more pressures on the natural environment. The islanders are also becoming more aware of their own challenges. The erratic global change and environmental challenges mean that the incidence of extreme rainfalls, soil erosion, coastal retreat, flooding, landslides and poor air quality in Taiwan is increasing. Here it is argued that the concepts embedded in geoparks become 21st century tools for environmental management that engages more stakeholders than ever before. With this bottom up approach, partially facilitated or supported by various levels of governments and academic society, Taiwan's geopark communities are full of people who show their love to care for their own land.

References

Forestry Bureau of Taiwan. (2018). *Nature reserves*. https://conservation.forest.gov.tw/reserve. Accessed August 11, 2018.

Lee, C. T. (2016). Landscape conservation in a dynamic geomorphology. In Chou and Lin (Eds.), *National development under climate changes* (pp. 353–372). Taipei, Taiwan: Societal Risk and Policy Center, National Taiwan University Press. (in Chinese).

Lee, K. C., Wang, S., Ho, L. T., & Chang, S. C. (2010). Community-based framework for planning geoparks. *Taiwan Forestry, 36*(5), 13–21.

Lin, J. C. (1999). *The beautiful landscapes of terrain*. Taipei, Taiwan: National Taiwan University. (in Chinese).

Lin, J. C. (2014). *The top ten geomorphological landscapes of Taiwan*. Taipei, Taiwan: National Taiwan University. (in Chinese).

Lin, J. C. (2015). *The inventory of and interpretation of landscape resources of Hu-Jin and Tongpan islets*. Penghu, Taiwan: National Scenic Area Administration of Penghu. (in Chinese).

Lin, J. C., & Su, S. J. (2013a). *The dynamic landscapes of Taiwan*. Taipei, Taiwan: National Taiwan University. (in Chinese).

Lin, J. C., & Su, S. J. (2013b). *Landscape conservation of Taiwan*. Taipei, Taiwan: National Taiwan University. (in Chinese).

Wang, C. Y. (2017). The updated regulation for the cultural heritage of natural landscape and monuments. *Communication for Landscape Conservation, 45,* 2–7. (in Chinese).

Wang, S. (1997). *Landscape conservation*. Taipei, Taiwan: Min-wen Publisher. (in Chinese).

Wang, S., & Ho, L. T. (1999). Introduction to landscape conservation techniques. *Communication for Landscape Conservation, 11,* 6–8.

Wang, S., & Lee, K. C. (2002). *Report for establishing geoparks and its environmental monitoring and management*. Taipei, Taiwan: Council for Agriculture Taiwan. (in Chinese).

Su, S. J., & Wang, W. C. (2013). *How to mobilize community for geoparks?*. Taipei: National Taiwan Normal University Press and Forestry Bureau, Taiwan. (in Chinese).

8 The Development of Taiwan's Geoparks

8.1 Networks

The concept of a geopark network in Taiwan mirrors the development of the international trend for the Global Geoparks Network. Currently exist three Regional Geopark Networks, a European Geoparks Network, an Asian Pacific Geoparks Network and a Latin America and Caribbean Network, all of which demonstrate the function of co-learning. These networking activities contain the organization of conferences, workshops, training courses, field studies, as well as the promotion of sustainable tourism or the production and presentation of local geo-products and, thus, are fulfilling the criteria of and showing the values of each geopark. Through networking, the Taiwan Geoparks Network benefits from the exchange of ideas and experience of the management and governance mechanism of all geoparks. Since 2011 the networking conference has been held twice a year at different geoparks in Taiwan (Fig. 8.1).

It has been found that the sharing of experience and management practices as well as exchange programs is a way of profound and effective learning for all. It has also been found that networking conferences provide opportunity for various local societies to become involved with national geopark activities (Lin 2014b). With the chances of visiting other geoparks, each group could learn from one another and become more confident about how to proceed to conserve their own environment and to involve in a locally appropriate way, while achieving the values of geoparks. People can formula new visions after visiting geoparks in other countries although they know that they will have to work distinctively and locally appropriate to develop and to manage their own geoparks.

Through a networking conference, the geopark community members could learn and benefit relevantly. Basically, the network conference provides a platform for different geoparks to learn from others. The geopark community members would have the opportunity to share ideas and practices of how they run and manage their parks. They may be able to compare and contrast the achievements of their own geoparks and other geoparks. Such comparing and contrasting is important, as human beings learn from difference. Further, with international geopark visit trips, geopark community members even could contrast geoparks with different contexts, which give a strong motivation for self-betterment. The motivation leads to self-empowerment through various ways of learning and practicing. Through the networking, all the people could then use various communication APP software to link to each other and stay in contact. Regardless the side effects of social media for mis-communication (Dey et al. 2018), it has been part of the networking life, and messages and issues could be brought up or shared instantly. Urgent matters among or within geoparks could get proper attention and matters could even be solved nicely. This is a new development. People can now share their activities more easily and efficiently at a minimum cost to achieve great benefit.

Promotion of geo-products is another achievement from the geopark network meeting, and for some geoparks it is significant. Because many geoparks are located in the semi-rural or peri-urban areas, geopark communities are partially dependent on agricultural activities for a living. Therefor the procures out of the land, which is often unique in certain ways, are much conditioned by its own environments, such as soil, water, climate and even slope. Therefore, the uniquely produced agricultural produces or fruits often come with exclusive flavors or characteristic features that could be representative of the local milieu (Chyi 2002; Liu 2013). Thus the promotion of geo-products is linked with the local environment and its stories, or even stories that could be linked to a greater tectonic or structural foundation.

When the geopark people meet, they share the techniques and methods of growing, planting and marketing of their unique products. It becomes a vigorous network to promote each other economically, although economy is not being the major concern of promoting geoparks. Some community members even learn from their Japanese geopark exchange trip about how geo-products could be re-created and

J.-C. Lin and S.-J. Su, *Geoparks of Taiwan*, Geoheritage, Geoparks and Geotourism,
https://doi.org/10.1007/978-3-030-04895-2_8

Fig. 8.1 A group picture of the 6th Taiwan Geopark Networking meeting in 2014

promoted through a thick description of the local stories of people and cultures, based on the geological environment (e.g., Su and Wang 2013). Japanese geoparks are a particularly useful model for Taiwanese geoparks. For instance, the invention and use of a walking human model, representing a basaltic rock and the hexagonal crystallization of basaltic rocks, makes it a fun and attractive scene. It not only introduces visitors to the geological features, but also the very dynamic and versatile content of the geopark (Fig. 8.2).

8.2 Re-thinking Earth Science for Society

Earth science is not a common and popular science among school students in Taiwan for many reasons. Earth science is not a course that is taught from a young age. In Taiwan, Earth science is only taught at the senior high school level as a mandatory course where every student has the chance to expose themselves to the knowledge of the earth planet. Even this mandatory curriculum earth science is usually not taught with an easy and light-weight fashion. Lack of fun and thus short of motivation becomes a common problem among students on earth science (Kuo 2012). Further, because it is taught at senior high school when student are older, the course becomes a relatively knowledge-oriented subject. Its knowledge- and scientific orientated tendency makes the subject short of relevance to the daily life of students. It therefore becomes a challenge and a window of opportunity for the promotion of geological knowledge to be elevated to a daily life and stake-holding orientation. The stake-holders are so diverse that geoparks involve multi-facet frontiers (Fig. 8.3) (Liu 1999, 2014b; Su 2017). In the case of geoparks of Taiwan, most initial community members who concern and participate, are adults who are often experienced enough to link how the environment relates to the functioning of living. They have high motivation of learning about what is involved surrounding their own life. Therefore in the promotion of geoparks earth science thus is made softened for the society. This also explains why the bottom-up mechanism of geopark organization is important for landscape conservation (Chyi 2002; Lee et al. 2010; Lin 2014a).

In a small country with geo-diversity, Taiwan's geopark development is uneven because all geoparks are different. Each has their own individual characteristics and approaches. There is thus a fundamental need to re-think the function of earth science education for each geopark and for the society as a whole. Knowledge of earth science helps us to know the limitations for development as well as the potential value for the human society whatever the geological or geomorphological background.

Because the sensitivity of all the areas in the geopark varies, geo-sites that are vulnerable and the special scientific and educational value should be conserved with special care.

Fig. 8.2 Model of hexagonal rock columns in Sanin Kaigan Geopark, Japan

They are often zoned with a core area. The areas right next to and adjacent the core zones are called buffer areas. With good understand of the geological conditions of a geopark and properly zoned areas in the geoparks, good management become possible (Fig. 8.4) (Lee et al. 2010). From assessment of nine geoparks it can be shown that there are four essential features. The individual geo-sites, sites of scientific interest, iconic viewpoints and museums or centers are usually linked, grouped into core zones and ideally protected by a buffer area or protected "setting".

In Taiwan the unique values and characteristics of prospect areas were assessed and preparations were made for landscape conservation initially and then for geopark status. Based on the long term monitoring and evaluations of the unique geological sites, each area has distinguishing features with scientific and environmental values. Therefore, the landscape and the sites of each geopark are being carefully evaluated for an extended period of time. Furthermore, they have close connections with the local life and livelihood.

The geoparks do not exist solely to conserve its geological features, but also to conserve the values and livelihood of the local society in a sustainable way. Eco-tourism, environmental education and sustainable local development are included, as they truly reflect the UNESCO values for global geoparks. In Taiwan, it is even found that the local living moral values and tenets could be realized and manifest in geopark practices (Liu 1999; Su and Lin 2014).

The Matsu geopark and its history will make a good case of understanding how society should be a good aspect for developing geopark. Black tourism, as a result of Matsu's militarily dominated society, can be a good venue for perceiving Matsu anew today, and it could be utilized as a means to remember and to convey to the next generations of the Matsu islets' history and environment. Matsu geopark has indeed a lot to show and to learn from due to its once isolation from the rest of the world. Because of the limitation to development during Martial Law periods, which was very much a result of the militarily confrontational situation between the Republic of China and the People's Republic of

Fig. 8.3 Stakeholders of Taiwan geoparks

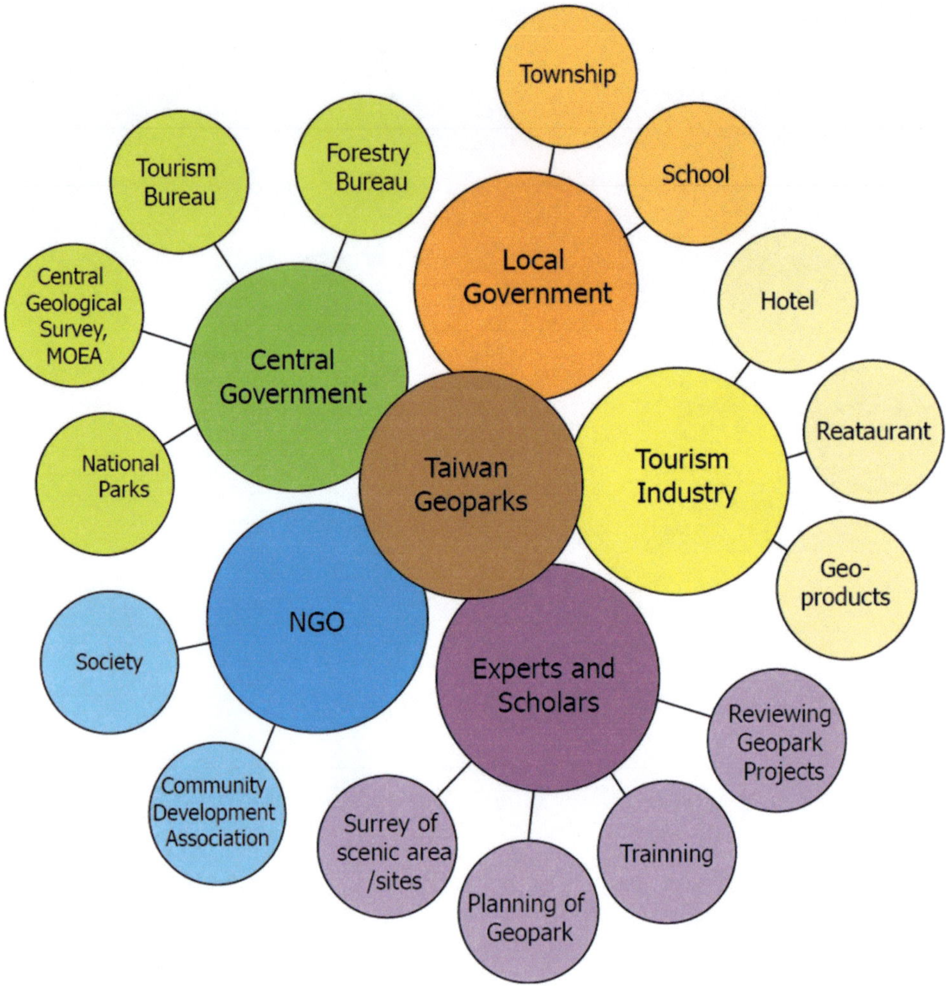

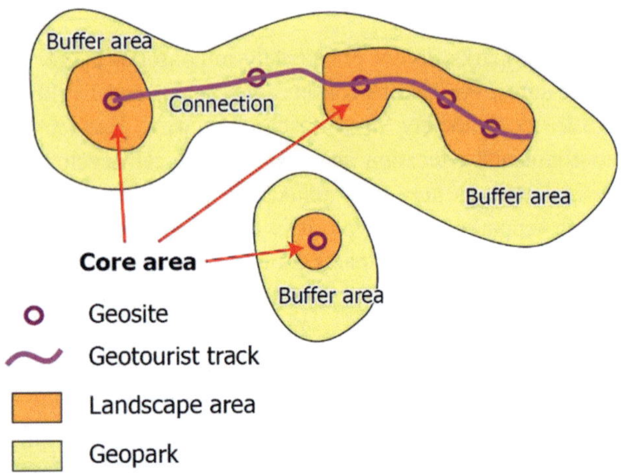

Fig. 8.4 A schematic diagram of a typical geopark with both core and buffer areas

China, Matsu's societal changes was kept slow during the last third of the 20th Century. There was a serious "lack of development" in this region although that might be beneficial for the islets in today's view (Su). However, the lack of the societal changes and perhaps lack of opportunity for individual or households caused by the special militarily administrative law has become a fact that brings the Taiwan society to re-evaluate the values of Matsu. Its unique architectures, special customs, authentic cruises and even the tranquil natural e2017nvironment reveal so much to be cherished today. The cultural and environmental values of Matsu can be well organized by its society of delicate and responsible civilians. Indeed, the local solidarity and bound of Matsu islet has been integral, which makes a strong community possible. The military history and its legacies can now become a unique potential for a new era of Matsu.

8.3 Bottom-Up Development and Networking for Geoparks

Typical cases where development is stimulated and pursued by the local community are Penghu, Yenchao, Lichi, Tsaoling and Matsu geoparks.

8.3.1 Penghu Marine Geopark

Various local groups and associations are zealously involved in local affairs, such as the Nature Study Club, and the Wild Bird Society-Penghu chapter. These groups were formed to appreciate and to protect the local unique milieu. For instance, the environment of Penghu has been an ideal habitat for migrant birds. Migrant birds such as terns come here annually for seasonal nesting and foraging which make bird watching activities highly cherished. As stakeholders, these groups are serious about nature conservation and environmental protection of the islets.

In early times, fishing was the main base economy in Penghu. Besides coastal, inshore and offshore fishing, many families are involved in intertidal stone weir fishing. That is why the building skills used to construct stone weirs are of great value. The famous Double-Heart Stone Weir in Cimei, coupled stone weirs of Erkan and the various fishing weirs of Jibei are good examples of the great traditional wisdom of local people in making good use of the local rock and coral materials at hand. Such development demonstrates the high utilization of earth science with traditional wisdom (e.g., Bureau of Cultural Heritage 2017; Liu 1999, 2015).

The remarkable local buildings are another example of how the traditional way of life and wisdom of the indigenous population can be used as a modern tourist attraction. Typical Penghu houses were built with local basaltic rocks and coral stones (Fig. 8.5). Houses like these are a combination of topographical features and human activities, not only requiring local techniques and aesthetics but practical use in a severe environment. The formal settlement in Wangan is one typical example. Wind breakers are needed for many activities in Penghu. Wind breakers are particular important for vegetables and settlements (Su and Lin 2014).

With mostly fishermen, local communities show great concerns about the marine environment (Fig. 8.6). Depending on the oceanic environment for fishes and sea weeds, the Penghu people are keen to use the knowledge and observation of marine environmental changes. The residents of the geopark communities understand the meaning of promoting geopark establishment. Some NGOs also play important roles of landscape and ecosystem conservation, such as the Nature Study Club, and Wild Bird Society on guiding eco-tour (Fig. 8.7).

Fig. 8.5 Traditional coral houses at Penghu Marine Geopark

Fig. 8.6 A fishing village at Tongpan islet, Penghu

8.3.2 Yenchao Geopark

Local participation and community networking is important for Geoparks. In the Yenchao Geopark, the Yuanjiao Humanist Association and some local village community associations are vital institutions active in promoting local beauty spot and networking with other geoparks. People combine community economy and badland characteristics to boost local development. Guavas and Jujubes are important cash crops and local cultural symbols. The activity of visiting the Funin- Jujube-Fields has become a core promotion activity for the community.

The Jingshan community also promotes its natural landscapes, traditional buildings, and cultural and agricultural economy through well planned activities, using traditional methods to make bricks, walls and even mountainous foot path for people who work in the badland fields (Chyi 2002). Jingshan is a place of hospitality and team spirit. The organization known as Granny's Rice Cake Team is zealous in promoting local recipes that use local produce. With its communal power, Jinshan has become the rear garden of Yenchao and Kaohsiung City. The best and most amazing of all these activities is that they make the best use of the elderly's wisdom and techniques to pass down virtues and values to the young generation.

The Yuanjiao Humanist Association promotes rural culture and history and has opened a new window to the discovery of the unique environments for Yenchao (Fig. 8.11). The association combines the cultural assets and natural beauties so that people are proud and willing to keep safe the local environment. Arranging activities related to the unique local culture helps people to understand their own culture and to conserve it.

Local communities can seek cooperation from many local organizations. When local people work together they create a novel community of networking to learn from one another. Examples are ample in the Yenchao Geopark community. The Yuanjiao Humanist Association, and the Jingshan Community association has proved to be vital institutions to achieve this. The locals from the Yenchao Bad Land and Mud Volcano Geopark portray themselves as the good people from the bad land. Viewing the barren land represented by the geopark's badland, one would have to ponder that yes indeed they are good people from the bad land. As the silt of the mud volcanoes can be detrimental to agriculture for their fine particles and thus the draining of water

Fig. 8.7 A guide of Penghu geopark conducts interpretation to visitors at Tongpan islet

is difficult. But with the local wisdom of irrigation system and pond system to prevent fine silts drain directly into the fields, the agriculture is developed well with wisdom (Fig. 8.8).

8.3.3 Lichi Geopark

Lichi is famous for its bad lands. Local demand for sustainable development is badly needed. The Lichi Badland Geopark is supported mainly by two communities, Fuyuan and Lichi each with their own cultural straits and identity and affection for their land. Participatory planning and workshops organized by experts and academics are keys to the promotion of this geopark. The geopark has been quite successful in incorporating neighboring communities and peoples of different origins to strengthen its geopark development. The concepts and values for geo-tourism and for the sustainability of the geopark can be a stimulus for revitalizing the aging and remote community in this beautiful part of Taiwan.

In the early stages of geopark promotion, the local community participation and partnership were critical. The geopark promotion activities also tried to incorporate elementary schools so that a scheme of environmental education for young people and local residents is made. With scientific geology, geomorphology, local economic activity and cultural life included, the program was board based and well integrated (Liu 2013). For example, for the community located on hills where Taiwan indigenous terraced agriculture is characteristic, the physical environment is dominant. For the Han Chinese people in the plains and coastal area, animal husbandry is a major economy as the soil is not as fertile and agriculture is not as easy. All these various activities involve very different viewpoints and practices. Environmental education at the geopark must therefore be integrated with the conservation values to conserve the use of the landscape by unique cultures.

Fig. 8.8 Mudstone "bad land" is challenging to agriculture

Muddy and easily eroded badland areas are not suitable for agriculture, but recently Lichi has developed many fruit orchards, with the policy support of the Farmers' Association from the Council of Agriculture. Among the well-known and popular fruits are guavas and sweet wax apples. Although Lichi is located in the eastern and remote area, its special fruits, landforms and ecology now appeal to crowds of visitors. The geopark promotion activities in recent years has not only increased the area's visibility, but also initiated a strong local response to the core values of geo-conservation. The rhythm and tempo of community life is closely accordant with its natural environment and the collective cultural/economic wisdom and way of life is loved by the visitors.

8.3.4 Tsaoling Geopark

Tsaoling, with varying geology and geomorphology, is located in the central mountainous Gukeng Township of Yunlin County and can be called a place with geo-diversity (Lee 2012). For its dynamic changing landforms and geomorphology, the geopark and its people has been facing great natural challenges (Lee 2013, 2014). Logging, agriculture and tourism are the main industries of the area (Lee 2011). With an average altitude of 800 meters, Tsaoling has broad-leaved evergreen forests and cedars. The local people use wood for decoration and handicrafts, and the forest adds a subtle misty beauty to the mountainous villages. Because of the moderate altitude, temperature, sunlight and other favorable physical conditions, Tsaoling is suitable for the growing of tea trees and coffee trees.

Tea trees are economically valuable because the tea oil extracted out of the tea tree seeds are in demand (Fig. 8.9). Due to its limited area of planting and its high technique-dependent manufacturing make edible tea oil an expensive product. Oil tea Camellia trees are pretty with blossom and attract tourists to enjoy the flowering scene before the seeds are harvested. Coffee trees here originated in the beginning of the 20th century. However it is only the last two decades when coffee drinking becomes a trendy practice in Taiwan, coffee making has found its niche in Gukeng and Tsaoling. The technique and methods for making fine coffee are pursued among the coffee growers in the area. That is one reason that the coffee produced here mostly sold out even before they have a chance to go on the

Fig. 8.9 Oil tea seeds bathing under the sun before extracting tea oil at Tsaoling

market. Such value-added aspect of tea oil, coffee and other produces makes the local people believe it is not only the techniques of making, but also the local milieu of water and soil that is critical to the quality of the produces.

Tsaoling has been infamous since 1870 for its huge landslide (Fig. 8.16) (Lee 2014). It is ironical that the fragile landscapes and local physical features became the foundation for attracting tourists. But such conditions are also where good soil and water occur and yield rich and expensive produce. The remoteness also makes Tsaoling a place of isolation. However the beautiful landscapes (Figs. 8.10 and 8.11), low population, low supply of human resource and little public infrastructure make Tsaoling a place of opportunity. The local society badly needs economic input from both outside and inside of the area. The local people have tried their best to run the place with energy. For example, the bamboo shelters and handicrafts are profit making ventures. These service tour buses with drivers who can interpret the landscape and this have become a strong attraction and a good source of income (Fig. 8.12).

The Shibi community has built over 20 ha of bamboo garden, called the "Wu-yuan-er-jiao bamboo canopy", using moso bamboos and dried bamboo leaves and stalks. Walking under the green canopy and within the bamboos, is a beautiful tourist experience (Fig. 8.13). The enthusiasm of local residents can be felt simply by witnessing how the locals introduce the green canopy to tourists. Throughout the building of this green canopy and arcades, local identity is reinforced by close relationships and mutual trust within this community. This is a deep and unexpected bonus from the geopark status.

The Bamboo grove is a major feature of the Geopark. Makino bamboos and moso bamboos are located everywhere, and the air is infused with their fragrance. Many local residents are skilled in making bamboo crafts and stone carvings. Crafts, such as bamboo containers, bamboo

Fig. 8.10 A panoramic mountainous landscape from Tsaoling Geopark

ladders, bamboo column beads, stone mortar, stone tables, stone benches, and stone walls, demonstrate great traditional handicraft techniques. Besides dried bamboo shoots for food (Fig. 8.14), jelly fig is also Tsaoling's famous specialty. These agricultural products are big selling points of the Geopark, and we label them as geo-products.

Tsaoling Geopark is a typical geopark merging out of the fear of hazards. Tsaoling was a small village but later it was transformed into a tourist attraction in 1976 due to a landslide that blocked a section of a river to form a lake. The lake become a boating site and the dip slope besides it became a slope for the courageous to walk up and down at 45° of slope without help (Fig. 8.15). Its fame rose nationally as one of the Ten Most Renowned Scenes of Taiwan.

However, among all the unique scenes in Tsaoling, only two survived the '921 earthquake' (of 1999) and, thus, its tourism declined. The community then started a reconstruction plan and succeeded in 2004 in the establishment of Tsaoling Geopark. The communities cooperate with local schools and utilize local ecological resources to develop environmental education programs and to sustain the cultural economy. Tsaoling Geopark becomes famous for its generosity and as a facilitator for geopark networking conferences (Fig. 8.16).

8.3.5 Matsu Geopark

Communities play an essential role in the establishment of the Matsu Geopark. As the martial law and the military administration prohibited civilian communal activities and civilian spirit, many villages in Matsu lack the experience in local initiative. With the auspices of the Matsu Scenic Area Administration, however communities are starting to take control of local affairs in a participatory fashion for the better (Lin 2013).

The Tae-ban community and its community association have mobilized residents to remodel some deserted traditional houses and public spaces to improve the local milieu. Such efforts display the potential of community making. Other significant communities, scattered over the five major islands of Matsu include Chin-bi of Bei-gan, Chong-liu and Lehua of Dongying, Fuzheng, and Dapu of Dongying. They have proved to be indispensable for community development and for the Matsu geopark.

Matsu is famous for its aquaculture, fishery food processing and winemaking (Fig. 8.16). With a good marine environment and rich marine resources, Matsu's fishery resources and major fish catches includes yellow croaker,

Fig. 8.11 Fluvial landscape at Tsaoling Geopark

white pomfret, shrimp, Poli fish and countless crustaceans, including Mytilidae and Foshou shells which have a high protein content and high economic value. Good use has been made of the ocean resources, for example aquaculture for yellow croaker, fishery food processing and rocky shore sea fishing activities (Fig. 8.17).

Tae-ban is the major economic centre of Matsu based on fishing. The arrival of the military in 1949, transformed the base economy from fishing and agriculture into commerce for serving the military. Later with the relocation of the county government and the withdrawal of the military administration, its economy transformed again. Tourism is now a vital element for the local community. The changing face of Tae-ban in fact reflects that of the whole area of Matsu. With the relaxation of military control, the small island society is changing rapidly. Faced with seemingly out of control changes of everything, quality knowledge-based eco-tourism has become the locals' pursuit. That makes geological knowledge and geopark in demand. For now, Matsu people believe that they are able to maintain its natural and human environments alongside eco- and geo-tourism with geopark promotion (Figs. 8.18 and 8.19) (Su and Lin 2014).

8.3.6 East Coastal Geopark (also Hsiaoyehliu or Fu-gan Geopark)

The East Coastal Range is one of the five mountain Ranges of Taiwan. It lies very close to the eastern seashore of Taiwan. Extending north from Hualien River to the estuary of the Peinan River in Taitung, the East Rift Range has a length of 150 km and a narrow average width of 10 km (Liu 2013).

Many uplifted ridges run east-west, en-echelon along the East Coastal Range, demonstrating the frequency of tectonic activities in the past. Volcanic agglomerates, mud stones, plentiful extrusive rocks and minerals present evidence that is used for geological and geomorphological research and environmental education. The area used to be famous for the excavation of gemstones.

The east side of the Range forms a spectacular coastline, where the sedimentary layers in Hsiaoyehliu and the raised coral reefs document environmental change. The Baxian Cave records the uplift of the range and erosion by seawater. The deposits also record the natural and human activities over the past 15,000 years and demonstrate the man-land relationships. The various coastal terraces and marine potholes at Shihtiping often attract visitors.

Fig. 8.12 Local geopark bus for tourists

Due to the moist air from the Pacific Ocean, the Coastal Range has a tropical rain forest biome with vines, ferns, epiphytes and other flora and fauna. Fuxing village at the foothills is a good example of how human society has adapted and taken advantage of the Coastal Range's environments since ancient time. The wisdom accumulated by the indigenous Amis people enabled them to develop sustainable livelihood under such conditions. This is a place where nature and human activities interact resulting in an intricate mix of cultural assets (Figs. 8.20, 8.21, 8.22 and 8.23).

8.4 Conclusion: Strategic Act for Promoting Geoparks ~ Taiwan Geopark Network Conference

Since 2005 when Taiwan organized its first geopark conference at Penghu, the development of a geopark network has been achieved gradually in many ways. In the beginning this was a bottom up social movement. Then the Bureau of Forestry promoted the geopark from the perspective of landscape conservation. Next, the Bureau of Tourism facilitated the concept from the perspective of tourism development. When Taiwan launched the Environmental Education Law, some efforts on environmental education became part of the mission in each geopark and local participation then became an essential component. From 6 geoparks in 2011 to 9 geoparks in 2016, there is now very strong local government support across the country according to the Cultural Heritage Law. All the geoparks have been re-evaluated and re-certificated as part of the Taiwanese geopark system.

The Taiwan Geoparks Network acts as a platform for each geopark to learn and share their experiences. The Network meeting also provides a mechanism for each geopark to learn from foreign geoparks, especially from Japan and Hong Kong. Since the first conference in Penghu, Taiwan geoparks also enjoy strong support from the international community and particularly benefitted from the experience of the Jurassic Coast in the U.K.

Fig. 8.13 The "Wu-yuan-er-jiao bamboo canopy" gives visitors the chance to stay and appreciate the tranquility of bamboo forestry

Fig. 8.14 Geo-products of bamboo from the Tsaoling Geopark

Fig. 8.15 Long and steep dip slope at Tsaoling Geopark

Fig. 8.16 Geopark networking conference at Tsaoling Geopark

Fig. 8.17 Laujeo (literally means *old wine*) is a typical rice wine from Matsu Geopark

Fig. 8.18 The explanation boards at Tae-bao geosite of the Matsu Geopark

Fig. 8.19 Explanation board for the turtle islet of Beigan at Matsu Geopark

Fig. 8.20 Coral capping a marine terrace represents sea level change during Holocene time according to the 14C dating

Fig. 8.21 Honeycomb textures on sandstone at Hsiaoyehliu Geopark

Fig. 8.22 Differential erosion landscape at Hsiaoyehliu Geopark

Fig. 8.23 The uplifted 3000–4000 years old sea cave at Bashendong on the eastern coast was once occupied by prehistorical population

Fig. 8.24 A scene from a welcome party held by Itoigawa Geopark of Japan for the Taiwan geopark delegates in 2014

In the future, each community of geoparks in Taiwan will manage their geoparks for their sustainable development. So far some geoparks, such as Yehliu geopark, even have 6 sister geoparks in Japan and China. Through regular activities among these geoparks, great efficiency is achieved (Fig. 8.24).

As a national agency promoting eco-friendly produces, the Bureau of Forestry provides a cloud of platform for each geopark to sell their geo-products on the website. All the geo-products have to be evaluated and quality-controlled to show that they support the health of the natural environment and support geopark societies.

All such developments demonstrate that the successful development of Taiwan's geoparks are mostly due to it societies and human endeavors that aspire to pursue sustainable way of life and in harmony with the nature. The man-nature relationship has become a special balance worthy of care.

References

Bureau of Cultural Heritages. (2017). Fishing weirs of Penghu as a potential site for world cultural heritage. https://twh.boch.gov.tw/taiwan/intro.aspx?id=13&lang=zh_tw#ad-image-0 Accessed July 2018.

Chyi, S. J. (2002). *The conservation guidebook of Tainan's geomorphological landscapes*. Kaohsiung, Taiwan: National Kaohsiung Normal University. (in Chinese).

Dey, N., Babo, R., Ashour, A. S., Bhatnagar, V., Bouhle, M. S. (Eds.). (2018). *Social networks science: Design, implementation, security, and challenges: From social networks analysis to social networks intelligence*. Berlin: Springer.

Kuo, C. C. (2012). The progress, limits and challenges of science education, Chinese. *Physics Education, 13*(1), 1–10. (in Chinese).

Lee, C. T. (2011). *The development process of Tsaoling Geopark*. Taipei: Forestry Bureau of Taiwan and National Taiwan University co-published.

Lee, C. T. (2012). *The geo-diversity of Tsaoling*. Taipei: Forestry Bureau of Taiwan and National Taiwan University co-published.

Lee, C. T. (2013). *The transformation of Tsaoling's geomorphology*. Taipei: Forestry Bureau of Taiwan and National Taiwan University co-published.

Lee, C. T. (2014). Landslides and Tsaoling Geopark. *Geology, 33*(1), 31–35.

Lee, K. C., Wang, S., Ho, L. T., & Chang, S. C. (2010). Community-based framework for planning geoparks. *Taiwan Forestry, 36*(5), 13–21.

Lin, J. C. (2013). *Introduction to Matsu Geopark*. Matsu, Taiwan: National Scenic Area Administration of Matsu.

Lin, J. C. (2014a). The ideal and practice of geopark promotion. *Geology, 33*(1), 6–8.

Lin, J. C. (2014b). Promoting Taiwan Geopark Network. *Geology, 33* (1), 20–22.

Lin, J. C. (2015). *The inventory of and interpretation of landscape resources of Hu-Jin and Tongpan islets*. Penghu, Taiwan: National Scenic Area Administration of Penghu.

Liu, S. L. (1999). *The cultural and tourist resources of Penghu Marine Geopark*. Penghu, Taiwan: Fishing and Agriculture Bureau of Penghu Government.

Liu, Y. S. (2013). *The environmental education guidebook for Lichi Badland Geopark*. Hualien, Taiwan: National Tunghua University. (in Chinese).

Su, S. J. (2017). Matsu's historical development and its implication for geopark development. *Geology, 36*(4), 55–62. (in Chinese).

Su, S. J., & Wang, W. C. (2013). *Manual for empowering geopark communities*. Taipei, Taiwan: National Taiwan Normal University Press. (in Chinese).

Su, S. J., & Lin, J. C. (2014). The environmental and social implication of geoparks. *Geology, 33*(1), 56–69. (in Chinese).

The Prospect for Geoparks in Taiwan

9

9.1 Evaluation and Assessment of Geoparks in Taiwan

The assessment of Taiwan geoparks follows the criteria and requirements of UNESCO Global Geoparks. This is underway as the geoparks are now part of the Cultural Heritage Act of Taiwan. All the existing geoparks and aspiring geoparks are now assessed according to the detailed and practical regulations and rules of the Cultural Heritage Act to be re-affirmed. After all the proper assessments and affirmations in accordance with the Cultural Heritage Act at each particular geopark are done, a county geopark or a national geopark status is recognized by the government.

The indicators used for this evaluation include: natural resources, local community involvement and governance, management units, managerial practice, and support from the local government and the relevant departments of central government. With some particular educational tools, environmental teaching curriculum and certain techniques and skills that have been co-developed by the academia and by the collaboration between the public and the private sectors (Wang and Ho 1999), the geoparks and their communities should very much have a good footing for further development.

The evaluation of Taiwan geoparks is carried out by a sub-committee of the national landscape evaluation committee (Lin 2014a). This committee has to help the geoparks to process for its design and plan for the evaluation. By law, there are two categories of geopark in Taiwan (Lin 2014b). The first category is the local geopark level with support from the local government and local community as well as local NGOs. The evaluation includes assessment of the natural resources and local participation in geopark activities. The geopark must have a set of landscape conservation data, environment education programmes, local participation of geo-products, geo-tours, geo-education activities, geo-heritage information and a geo-hazard resilience plan.

The second category is at the national level. If the geopark passes the evaluation and all the requirements by the central government, it will be classified a national geopark. A special requirement is that the natural resources must be of scientific and of international importance. Those national geoparks also need a strong management body to maintain the geoparks in good shape. The support of the local government is also essential. As the management of geoparks also needs funding for the activities and attendance at network meetings, raising funds by application to different departments of government is also important.

The evaluation of the value of the geopark for landscape conservation follows from the procedure of surveying the national scenic and geologically significant spots. Most of the geoparks have some hot spots/sites which attract visitors, especially those geoparks within the National Scenic Area. Those hot spots also provide resources for local sustainability.

The evaluation process includes assessment of:

1. Resource surveying of the geopark by an NGO, university or government department
2. Educational programs and capacity building for the local community
3. Activities and management
4. Proposals from the local government
5. Public discussion of the proposal
6. The agreement to the proposal by the committee organized by the County government
7. The agreement to support the local geopark by the County government, after evaluation
8. The management of the geopark with designation of a clear core area, a buffer area as well as the local involvement and protection management programs for the site
9. The planning for sustainability status of geopark
10. Provision of training courses for the guide tours and geo-products

J.-C. Lin and S.-J. Su, *Geoparks of Taiwan*, Geoheritage, Geoparks and Geotourism,
https://doi.org/10.1007/978-3-030-04895-2_9

11. Relationships between the management and the central government
12. The evaluation by the central government on all the requirements (Fig. 9.1).

9.2 Management of Geoparks

As a geopark is a formal institution that involves both human and non-human agents, its management and governance might be complicated. However, from the human agency perspective the management of geoparks can be categorized into three frontiers in Taiwan's experience of geopark development (Fig. 9.2):

Fig. 9.1 The procedures for becoming a Taiwan Geopark

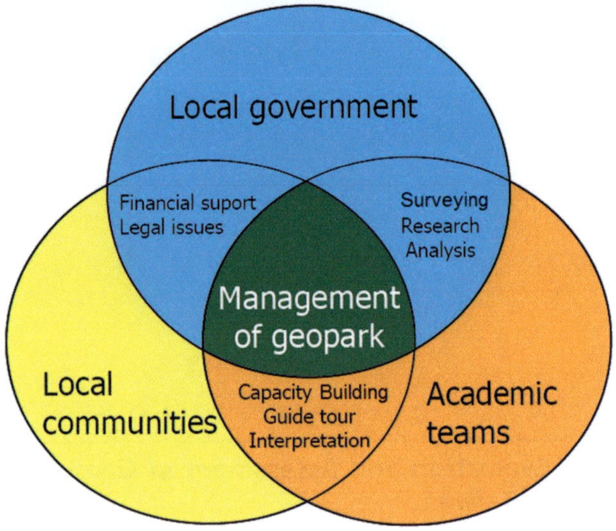

Fig. 9.2 Managerial scheme of Taiwan's geopark from a human agency perspective

1. Local government and other department of governments, such as the National Scenic Area Authority: Those are mainly sources of funding. As the geoparks are located in scenic areas, the authority is responsible for the management of facilities, such as transportation, explanation plates and other information facilities.
2. Local communities: Local communities act as the local platform for the governments to assist the management. Some geoparks provide free guides, such as the Yenchao or Yehliu geoparks. They also produce a regular series of publications.
3. Academic teams: The academic teams help local government to carry out surveying of natural resources and publications, such as geopark guides, pamphlets and information on websites. The academic teams also call together network meetings usually twice a year and a workshop every year. The academic team plays a more active role when the geopark concept is introduced at the beginning of the establishing process (correct 'Surveying' in Fig. 9.2).

Most geoparks are at the beginning stage for management. There are some geoparks located within national scenic areas, where there is an official authority which is responsible for overseeing its progress and development, such as the geoparks at Penghu, Matsu, Yehliu, Bitou-Longdong, East Coast and Southwest Coastal wetland geoparks. The Yenchao, Tsaoling and Lichi geoparks are mainly governed by the local communities with some support from Forestry Bureau. In addition, some aspiring geoparks are located within the boundary of National Park authority. As geology has been one important foundation for geoparks, the Taiwan's Central Geological Survey has been crucial as well.

The geoparks within the National Scenic Areas normally provide tourist information centers with regular publications, guide tours by volunteers and multi-media presentations. Interpretation and explanation boards are also ready at most of the scenic sites. These geoparks normally have regular budgets for management. Other geoparks are still at the beginning stage and need more support from the local governments. Such differences reflect the variability in the level of management of geoparks in Taiwan.

9.3 Aspiring Geoparks in Taiwan

According to the landscape surveying program and the Cultural Heritage Act, there are many other potential sites which could be designated as geoparks. Based on geological and geomorphological resources, some sites, especially at many national parks, together with activities of local communities, are promising geoparks for the near future. Examples are:

Yangmingshan National Park

Northern Taiwan is largely made up of volcanic landscapes. Presently the Yangmingshan area, located close to the largest metropolis of Taiwan, is designated as Yangmingshan National Park. The national park is famous for its volcanic landscapes characterized by the Datun Mt. volcanoes, which are mainly composed of andesite. Unique conical and bell-shaped volcanoes, volcanic craters and crater lakes all have extraordinary geological and geographic values. As these are all valuable scenic sites for the urban population, their geological values and their implications related to geoparks are yet to be further explored by the local community and visitors (Figs. 9.3, 9.4 and 9.5).

In this area, surface water infiltrates the underground and is heated by geothermal flux, resulting in fumaroles and

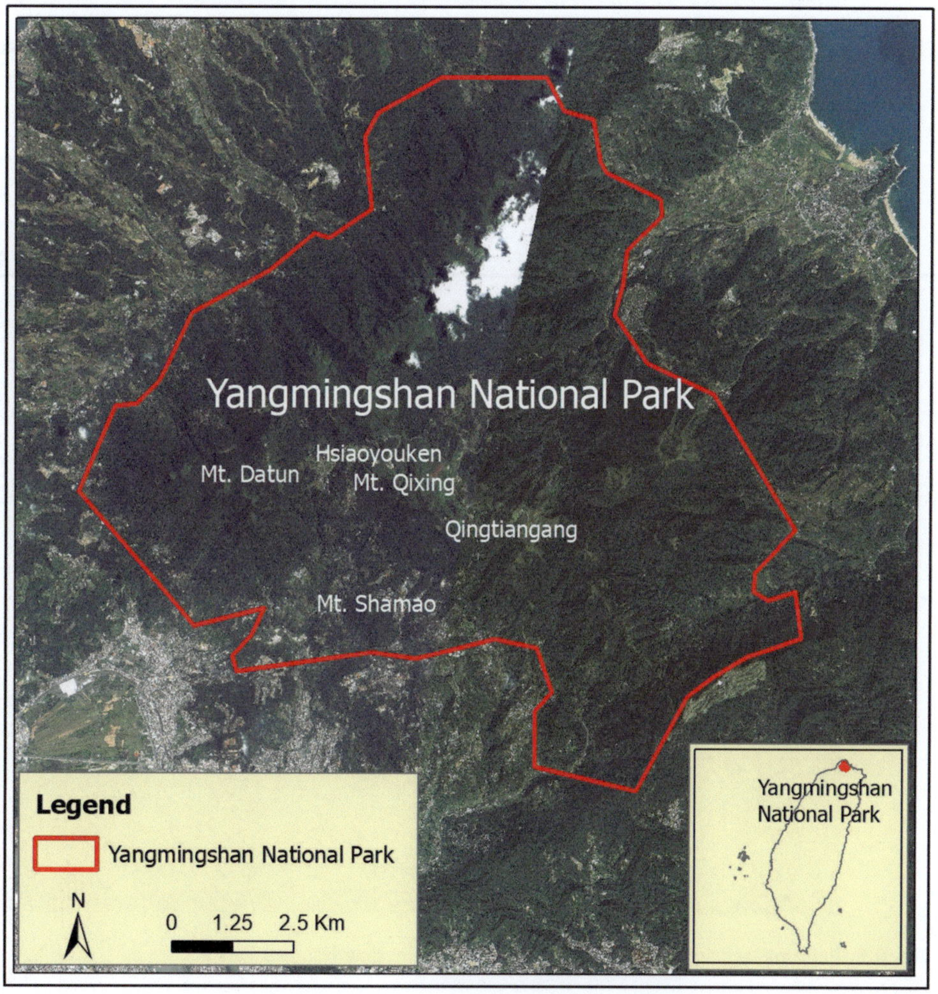

Fig. 9.3 Location map of Yangmingshan National Park

Fig. 9.4 A scene as a result of fumarole activity, hydrothermal alterations and weathering of volcanic rocks in Yangmingshan area

Fig. 9.5 Fumarole activity at Xiaoyoukeng in the Yangmingshan National Park

thermal springs (hot springs). These volcanic activities are part of the environmental resources that influence the characteristic human daily life of the area. The seismo-tectonic activities are causing frequently structural and fault movements of the island and contribute to an increased vulnerability of the society, thus, the area's physical foundation deserves more attention by the society.

Yangmingshan was once called Grass Mountain, located close to and north of the Taipei basin. The history of human settlement can be traced back to at least 2000 years ago when the Ketagalan people settled, made a living and stamped their culture on the area. Various cultural heritages of the Dutch, the Spanish and even the Japanese have been identified by careful observation. Today, the area is a popular resort for the urban population for weekends and public holidays. The interesting mix of past cultures and settlements has resulted in an important archive for present and future generations to cherish.

The mineral deposits formed by the fumarole activity, hydrothermal alterations and weathering of volcanic rocks in the Yangmingshan area include sulfur, pyrite, limonite, acristobalite ("white earth"), hokutolite deposits and bauxite. Today, mining activity takes place only on a very small scale mainly involving sulfur and white earth. This clay was once mainly used as ingredients of making pottery and building and today it is used with more added value as chemical ingredients for fertilizers and other modern industry.

Besides the hot springs areas exist where great clouds of geothermal steam rush out of fumaroles of all sizes. These are encrusted with yellow sulfur crystals. Other typical geological features of the Datun volcano group are crater lakes and geological faults. Volcanic cones are formed by the piling up of lava flows and pyroclastic rocks around the main volcanic craters. Mt. Qixing is a good example of this type. At an elevation of 1120 meters above sea level, it is the highest peak in the Yangmingshan National Park and in northern Taiwan. It is one of the younger volcanoes in the park, and it has the most intact conical shape. Volcanic activity in the Yangmingshan area tailed off after the formation of Mt. Shamao some 300,000 years ago, leaving the various signs of post-volcanic activity we see today.

Yangmingshan's popular hot springs can be divided into three types according to their chemical composition (Song 2005):

- Acid sulfate (white sulfur) hot springs, at Dahuangzui, Zhuzihu, Sihuangziping and Gengziping areas.
- Acid sulfate chloride (green sulfur) hot springs, mainly at Xinbeitou area.
- Neutral iron-bearing hot springs (white sulfur), cooler and mostly consist of indirectly heated groundwater, located mainly at Dingpi Bridge and Lengshuikeng areas (Table 9.1).

The purpose of establishing a national park is to protect the unique natural landscape, native species, cultural and historical sites, as well as provide a place for leisure, education and research. The area was originally established as a protected habitat with the aim of conserving species and restoring the ecosystem. The Yangmingshan National Park, for example, has an average of almost 13,000 visitors a day, with seasonal variation in which flowering March is the busiest month. In general, the number of visitors can reach more than 35,000 $\sim$ 50,000 on a fine spring day. Such a large number of tourists impact the park's natural and cultural resources. The priorities of a national park are not the same as geoparks which have a social and economic function for the community. Without proper management schemes, assets could turn into liability. Nevertheless, it is possible to include geopark areas within national parks. The Yangmingshan area could be a good exemplar where the pragmatic values and methods of the geopark could relieve pressure from the National Park resources (Fig. 9.6).

Taroko National Park

The Taroko National Park is famous for its marble gorge. The Liwu River is the driving force and together with active tectonic processes they have shaped this dynamic landscape (Wang 2016). As the exposed marble is hard and resistant, with river erosion and uplift of the continents occurring at

Table 9.1 Characters of hot spring around Yangminshan National Park

	Site	T(°C)	pH	SO_4^{-2}	Na^+	K^+	Mg^{+2}	Ca^{+2}	Fe^{+2}	Al^{+3}	Si^{+4}
1	Liouhuang Valley	50–97	2.7	500	7.44	4.88	4.80	21.0	13.0	68.0	103
2	Jhongshan Building	45–65	2–4	278	13.0	5.27	15.2	24.8	3.65	22.1	111
3	Xiaoyoukeng	65–99	2–4	1015	96.9	18.2	85.6	381	47.3	37.2	209
4	Macao	60–90	2–4	346	16.2	6.02	19.0	55.9	9.54	9.38	152
5	Dayoukeng	65–98	1–2	225	14.8	1.50	2.30	7.10	4.00	1.50	9.80
6	Bayan	60–85	2–3	344	78.9	47.9	48.4	86.8	55.3	73.2	188

Fig. 9.6 The Hsiaoyouken visitor center at Yangminshan National Park

the same time, very deep narrow gorges are formed. Geologically, four million years of uplift, erosion, deposition and fluvial erosion have crafted the vertical valley landscapes of Taroko (Fig. 9.7).

The dynamic orogenic forces have made this area an excellent witness of the dynamic landscape of planet Earth. From high mountains to sheer gorges, from erosion, transportation to deposition, they all demonstrate the scientific principles that underpin the shaping of landforms. With the addition of a unique human culture and diverse economic activities, this site is surely an excellent geopark candidate.

The area of the Taroko National Park has abundant historical monuments and relics, such as prehistoric sites, the Taroko culture, and the ancient trail systems. For example, the Psngan Ruins, located close to the estuary of Liwu River is a national historical relic. The river terrace site used as a settlement can be traced back to the Neolithic Age, dating back 2000 years. Its inhabitants were the ancestors of the Taroko people, who reportedly migrated from the upstream of the Zhuoshui River to this area. The archeological evidence tells of the ancient settlements and their relationships and currently 79 archeological sites are under study. The deep and narrow gorge presented a difficult way of life for the ancient population. That resulted in the numerous human relics.

From the dynamic natural landscape, the extraordinary ecology, and the astonishing historic monuments of human activities, the Taroko area with its ethnic population also represents an ideal geopark. Furthermore, during the construction of the highway through the Gorge in the 1970s, some veteran workers sacrificed their lives in the strenuously condition of the gorge area. Evergreen Shrine is thus a memorial temple that was built for commemorating these people. Today as tourists pay visits to the Gorge area, safety and good interpretation of its historical, cultural and natural history is important and could be provided by the Taroko people (Figs. 9.8, 9.9 and 9.10).

Yushan National Park

Taiwan's high mountains have elevations generally between 3000-3500 meters. The variations in the elevations today are the result of differential erosion. These mountains were formed during the Miocene epoch, dated from 23 to 5.3 million years ago. During this period, warmer global climates than those in the preceding Oligocene or the following

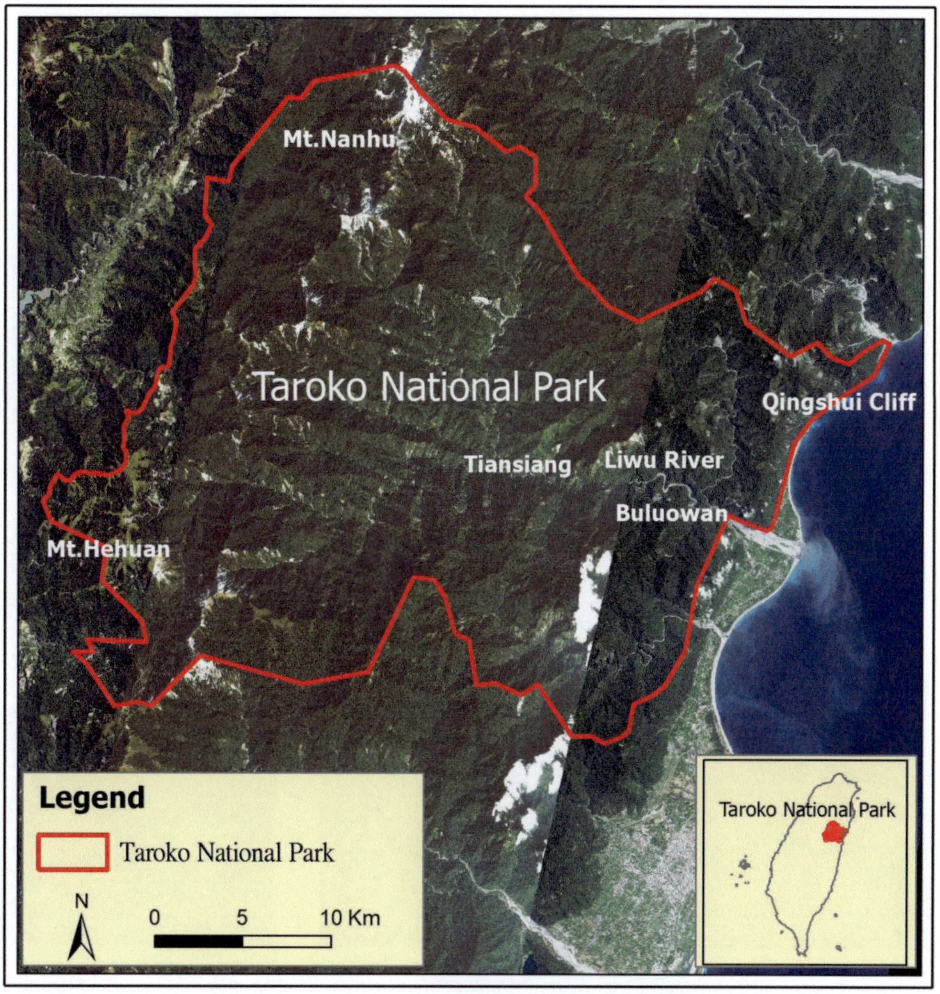

Fig. 9.7 Location map of Taroko National Park

Pliocene existed, and two major ecosystems, the kelp forests and grasslands, made their first appearances. The expansion of the grasslands is correlated to the drying continental interiors as the global climate first warmed and then cooled. The rocks of the area portray these changes. There are more than two hundred peaks higher than 3000 meters in Taiwan. The major vegetation type of these high mountains is alpine meadows as the altitude is above the forest line (National Museum of History 2004). Scattered gravels are the evidence of severe physical weathering (Figs. 9.11, 9.12 and 9.13).

Alpine landscapes are commonly found in Taiwan's mountains (Lin 2012, 2014c). Fluvial erosions cause deep valleys due to down-cutting. In spring time, flowers, alpine grasslands and trees are integrated into a delightful landscape enjoyable for tourists. Alpine meadows are mainly found around mountain areas of 3500 meters high. The simple straight lines of alpine meadows, with long slopes and the silhouette of bamboo clusters, show the delicate beauty of Taiwan's wilderness. A wonderful example is the alpine meadows of Houhuan Mountain. Unfortunately, roads and other constructions often interfere and degrade such natural beauty and, thus, selected protection is urgently needed to allow the establishment of a geopark (Figs. 9.14 and 9.15).

Within the Yushan National Park boundary, there are two major indigenous villages, namely Dongpu and Meishan, inhabited by the Bunun people. The Bunun were originally residents of Taiwan's western plain. Decades of Han Chinese encroachment and the search for better hunting grounds forced the people further eastward and into the central mountains of Taiwan. The Bunun people are critically important in safeguarding the Yushan mountainous area. Because it is a place upon which their livelihood is dependent, and in which the bunun's scared culture and folklores are preserved. For the safe keeping of the Bunun heritage, conservation with traditional wisdom by the Bunun people becomes imperative.

Fig. 9.8 Faulted cliff and spectacular road of Taroko National Park

Fig. 9.9 Evergreen Shrine as a memorial site

The indigenous word of Bunun means "people". It was a self-claiming term of the Bunun people to distinguish themselves from other peoples. They inhabit the southern part of the Central Range with a population of approximately 40 thousand, and around half of whom dwell in mountainous areas at the altitude of 1500 meters. A typical mountain people, the Bunun is one of the most energetic, wide spreading, and the best adapted for the mountain environments among the natives of Taiwan. The Bunun traditionally conducted slash and burn farming for sustenance, supplemented by hunting and the gathering of food plants from the wild. Their early distribution was mostly along the terraces of river banks, which played an important role in the Bunun culture.

The folk songs of the Bunun are one of the treasures of ethno-musicology in the world. In their pure voice, some amazing choral of contrast and harmony can be enjoyed. Since the Bunun singing is part of thier festival and celebration, and integral with agricultural production, one can feel their life energy upon hearing them roaring and whistling in the mountains. This will certainly add one unique feature to the Taiwan geopark with vocal presentation surrounding the mountains and valleys.

Kinmen National Park

The Kinmen National Park is famous for its war and military heritage (Chiang 2002, 2005). The heritage was from the cold war between Taiwan and China between 1945 and the early 1990s. Kinmen National Park still keep many old vernacular housing architectures which reflects the societal development before the 1950's. Now some of the houses have been renovated and preserved as historical and cultural heritages. With its status of being a military controlled islet group, Kinmen was kept isolated and many socio-cultural traditions and architectural features are maintained perfectly intact (Chiang 2014) (Figs. 9.16, 9.17 and 9.18).

Fig. 9.10 Gorge landscape at Taroko National Park

Geology-wise, Kinmen is mainly composed by granite landscapes. Due to differential erosion, its geo-diversity and geomorphology is rich especially at all the headlands of Kinmen. For defense purposes, these headlands were designated with strongholds and military facilities. Because such military structures were not allowed to be accessed by the public and thus with little influence of human activities, these headland areas are great scientific laboratories for geology, geomorphology and ecology.

Among all the war and military artifacts the anti-landing facilities haves become tourist attractions. The anti-landing facilities along the coasts of Kinmen were build up in the cold-war-times due to the close proximity to China which was considered a threatening and confronting country. Fortunately, that period is over, the military constructions ended up becoming coastal environmental threats. Turning the anti-landing facilities into educational materials to explain the military history and the cold war could be part of the geopark values of cherishing the environmental past (Figs. 9.19 and 9.20).

As Kinmen's cultivating and agricultural environment was infertile and even poor at some places, the livelihood of people was limited. In addition, endangered by the coastal location in the monsoon directions, emigration to other countries of Southeast Asia like Indonesia, Malaysia and the Philippines was a solution for feeding large family populations. Therefore, the Kinmen Island was once called the hometown of a Chinese diaspora. This history caused the Kinmen people to cherish family ties even greater. Those Kinmen migrants living abroad in foreign countries often send remittance home for buildings to house family members. Therefore, today's Kinmen vernacular architectures present some features and elements that are foreign or even clearly Southeast Asian in origin (Fig. 9.21). With their rich socio-cultural, diasporic historical, island ecological and geological significance, Kinmen islets can add to Taiwan Geoparks with additional value and beauty of trans-cultural diversity.

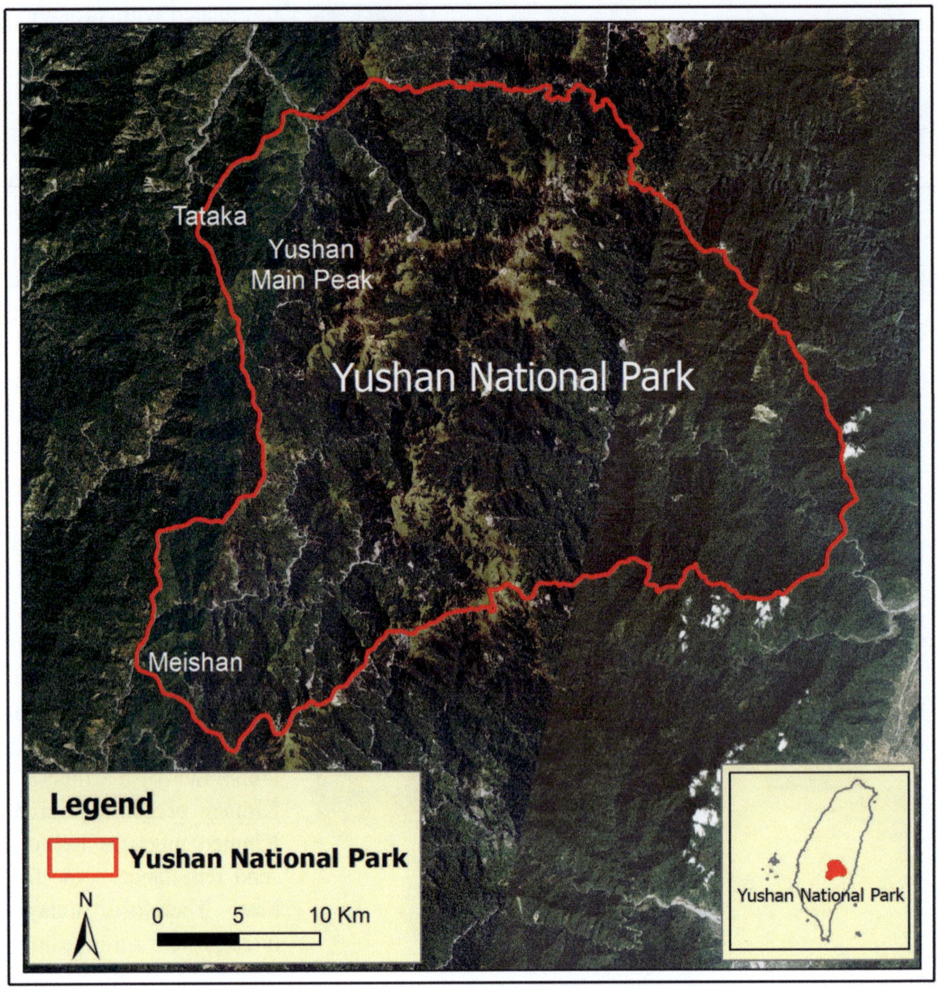

Fig. 9.11 Location map of Yushan National Park

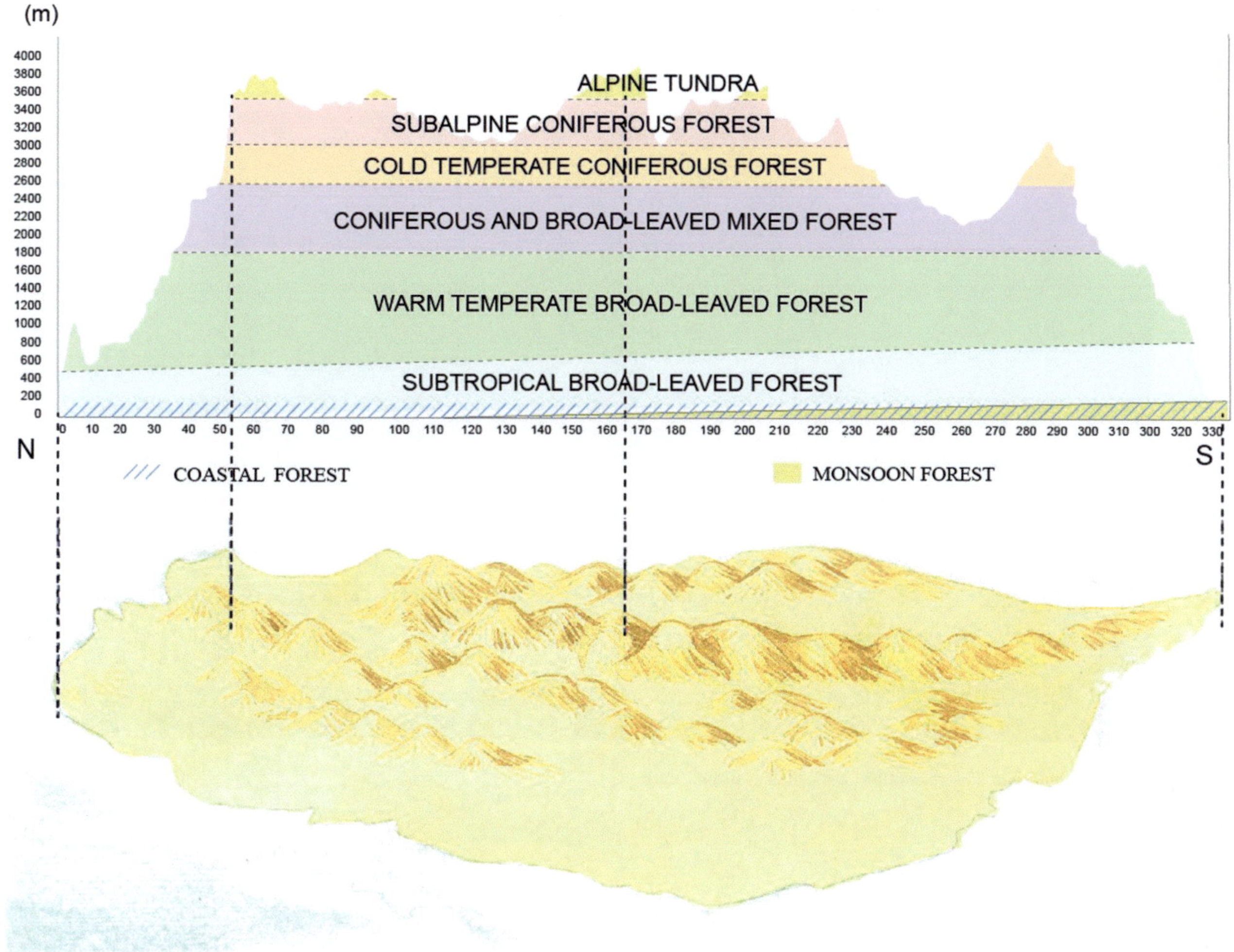

Fig. 9.12 Vegetation distribution at different heights in Taiwan

Fig. 9.13 Alpine landscape and vegetation at Yushan National Park

Fig. 9.14 The highest peak of Taiwan, Yushan National Park

Fig. 9.15 Trail with hikers at the fragile landform of Yushan National Park

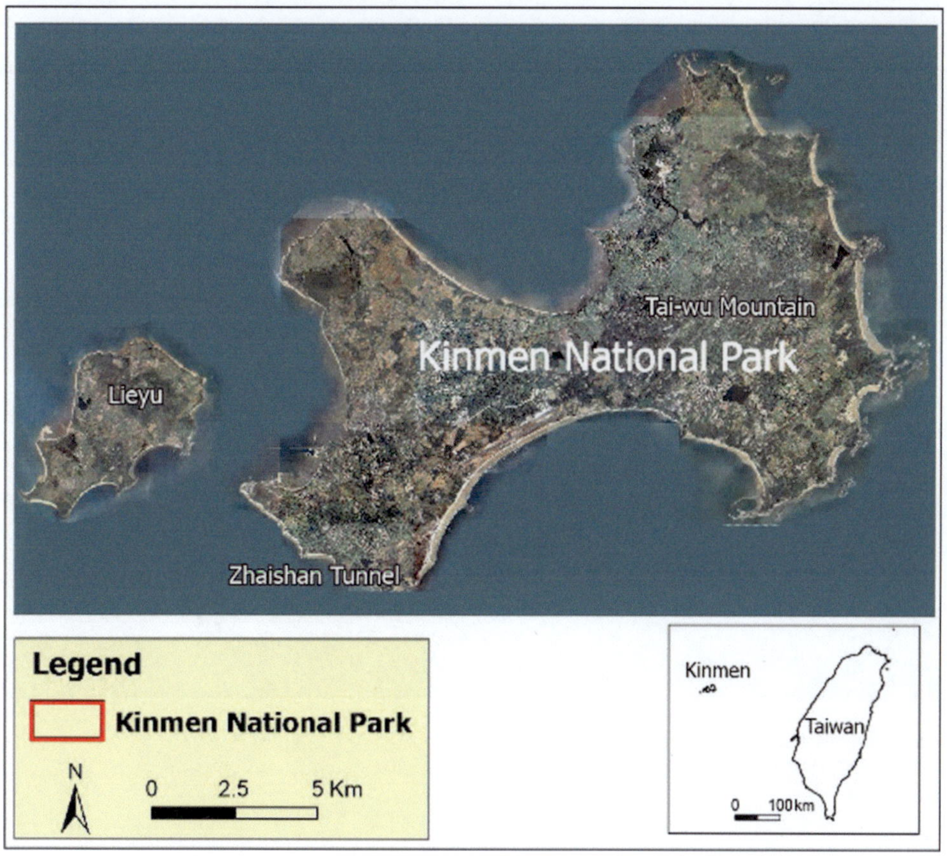

Fig. 9.16 The location map of Kinmen National Park

Fig. 9.17 Marine terrace of iron-rich sandstone at Leiyu, Kinmen National Park

Fig. 9.18 Intact historical tower built upon granite in Kinmen

Fig. 9.19 An artificial tunnel dug during the cold war era

Fig. 9.20 Anti-landing facility to prevent enemy's attack as a cold war heritage

Fig. 9.21 Architecture of Southeast Asian influence revealing Kinmen's migration history

References

Chiang, B. W. (2002). *The Ju-Kuang tower at Quemoy: Its genius loci and military nationalism*. Kinmen: Kinmen County Government Press.

Chiang, B. W. (2005). *The Almanac of military confrontations in Quemoy*. Kinmen: Kinmen County Government Press.

Chiang, B. W. (2014). *Transforming Quemoy: Its vernacular architectures and almanac of Changes*. Kinmen: Kinmen County Government Press.

Lin, J. C. (2014a). The ideal and practice of geopark promotion. *Geology, 33*(1), 6–8.

Lin, J. C. (2014b). Taiwan's Geoparks. *Geology, 33*(1), 18–19.

Lin, J. C. (2014c). *The Top ten geomorphological landscapes of Taiwan*. Taipei, Taiwan: National Taiwan University.

Lin, T. C. (2012). *The great geological exploration of Taiwan*. New Taipei City, Taiwan: Taiwan's Central Geological Survey.

National Museum of History. (2004). *An epitome of Taiwan: The Jade Mountain*. Taipei, Taiwan: Editorial Board of the National Museum of History.

Song, S. R. (2005). *The inventory of hot spring and thermal heat in Yangmin Mountain*. Taipei, Taiwan: YMS National Park Service, Planning and Construction Agency, Ministry of Interior.

Wang, S. (2016). *The stories of Formosan Landscapes: Its unique features*. Taipei, Taiwan: Yuan-chu Publisher.

Wang, S., & Ho, L. T. (1999). Introduction to landscape conservation techniques. *Communication for Landscape Conservation, 11*, 6–8.

MIX
Papier aus verantwortungsvollen Quellen
Paper from responsible sources
FSC® C105338

If you have any concerns about our products,
you can contact us on
ProductSafety@springernature.com

In case Publisher is established outside the EU,
the EU authorized representative is:
Springer Nature Customer Service Center GmbH
Europaplatz 3, 69115 Heidelberg, Germany

Printed by Libri Plureos GmbH
in Hamburg, Germany